When we think of innovation, we often think of scientists
and engineers toiling away in secretive R&D labs.
Not so at Intel.
Will.i.am, star of the Black Eyed Peas band, was named
Intel's Director of Creative Innovation.

Indeed, something has changed
in the world of business innovation—changed utterly.
And it's not just where ideas come from,
it's how the work of innovation gets done;
it's all about *execution!*
This book is about *executing* on innovation
in the era of cloud computing.

Will.i.am

—Early Comments—

Absolutely nothing in the cloud computing "stack" of technologies is new: virtualization dates to the 1960's, networking almost as far (and certainly well-established by the beginning of the millennium), even small servers ("blades") and co-location racks have been around. What's interesting, what allows for disruptive innovation, is the convergence of cheap virtualization, cheap (and high-bandwidth) networking, and cheap racks of blades—and the recognition that they can be rented. As Stephen Udvar-Hazy, CEO of Air Lease Corp., taught the airlines, you don't have to own your most visible assets. Once you understand that, how do you innovate? Converting capital expenses to operating expenses is just the beginning. Thinking differently about computing expenses, location and architecture opens new vistas to see. Integrating crowd-sourced and customer-sourced innovation has never been so straightforward as in the new Cloud world.

Stikeleather and Fingar hit a home run with this ode to technology and business convergence, and given the innovation imperative, make this the first book to turn to on business innovation in the Cloud.

Dr. Richard M. Soley
Chairman and Chief Executive Officer
Object Management Group, Inc.

Think you understand how technology is going to change the way we work, organize, and compete? Think again. In this exciting and deeply insightful book, Stikeleather and Fingar offer a compelling vision of how the Cloud will rewrite the rules of business. For them, the Cloud isn't about "on-demand IT," but about "on-demand innovation"—and if your organization doesn't get this, it's gonna get caught out by the future. In a book that is both practical and thought-provoking, the authors do a great job of unpacking "the next big thing."

Dr. Gary P. Hamel
London Business School
Author of *What Matters Now*

Business Innovation in the Cloud

Look for periodic updates at
www.mkpress.com/BIC

Companion Books

DOT CLOUD:
THE 21ST CENTURY BUSINESS PLATFORM BUILT ON CLOUD COMPUTING TECHNOLOGIES

ENTERPRISE CLOUD COMPUTING:
A STRATEGY GUIDE FOR BUSINESS AND TECHNOLOGY LEADERS

BUSINESS ARCHITECTURE:
THE ART AND PRACTICE OF BUSINESS TRANSFORMATION

MASTERING THE UNPREDICTABLE:
HOW ADAPTIVE CASE MANAGEMENT WILL REVOLUTIONIZE THE WAY
THAT KNOWLEDGE WORKERS GET THINGS DONE

EXTREME COMPETITION:
INNOVATION AND THE GREAT 21ST CENTURY
BUSINESS REFORMATION

BUSINESS PROCESS MANAGEMENT:
THE THIRD WAVE

IT DOESN'T MATTER:
BUSINESS PROCESSES DO

THE REAL-TIME ENTERPRISE:
COMPETING ON TIME

THE DEATH OF 'E' AND
THE BIRTH OF THE REAL NEW ECONOMY

Acclaim for our books:
Featured book recommendation,
Harvard Business School's *Working Knowledge,*

Book of the Year, *Internet World*

Meghan-Kiffer Press
Tampa, Florida, USA
www.mkpress.com
Innovation at the Intersection of Business and Technology

Business Innovation in the Cloud

Executing on Innovation with Cloud Computing

Jim Stikeleather
Peter Fingar

Meghan-Kiffer Press
Tampa, Florida, USA, www.mkpress.com
Innovation at the Intersection of Business and Technology

Publisher's Cataloging-in-Publication Data

Stikeleather, Jim and Peter Fingar.
Business Innovation in the Cloud: Executing on Innovation with Cloud Computing / Jim Stikeleather, Peter Fingar, - 1st ed.
p. cm.
Includes bibliographic entries, appendices, and index.
ISBN-10: 0-929652-18-5 ISBN-13: 978-0-929652-18-4
1. Management 2. Technological innovation. 3. Diffusion of innovations. 4. Globalization—Economic aspects. 5. Information technology. 6. Information Society. 7. Organizational change. I. Stikeleather, Jim. II. Fingar, Peter. III. Title

HM48.S75 2012 Library of Congress No. 2012932911
303.48'33–dc22 CIP

Published by Meghan-Kiffer Press
310 East Fern Street — Suite G
Tampa, FL 33604 USA

Meghan-Kiffer Press
Tampa, Florida, USA
Innovation at the Intersection of Business and Technology

Printed in the United States of America. SAN 249-7980
MK Printing 10 9 8 7 6 5 4 3 2 1

Table of Contents

FOREWORD BY DEREK MIERS, FORRESTER RESEARCH 10

NOTES 12

PREFACE 14

PART 1. THE STATE OF BUSINESS INNOVATION 16

1. PROLOGUE: SOMETHING IS HAPPENING 17

EVERYTHING 2.0 17
WHEN MEMORIES EXCEED DREAMS, THE END IS NEAR 20
WHAT IF? 22
QUESTIONING INNOVATION 27
TAKEAWAY 30

2. DO THE WORK 32

TAKEAWAY 37

3. CHANGE ITSELF IS CHANGING 38

MACRO TRENDS AFFECTING THE WORKPLACE 42
Work Mobility and the Hollywood Model 43
A New Workforce / An Old Workforce 43
The New Normal: IT Consumerization 46
The New Normal: Unexpected Economic Change and Innovating to Zero 48
Cloud Computing: A Technology-Enabled Game Changer 49
Cloud Computing: A Business Model Game Changer 52
Risk and Security Management: Regulations and Cyber Jurisprudence 54
Smart Everything + Pervasive Simplicity = Massive Computing Needs = The Cloud 56
The Cognitive Computer 57
Windows to the One 59
TAKEAWAY: PUTTING THE MACRO TRENDS ALL TOGETHER 62

4. WHAT EXACTLY IS BUSINESS INNOVATION? 63

WHERE DOES INNOVATION COME FROM? 69
CREATIVE DESTRUCTION 2.0 71
CREATIVE DESTRUCTION 3.0 79
TENETS OF INNOVATION 86
INNOVATION'S RELATIONSHIP TO CREATIVITY 88
THREE CORE TYPES OF INNOVATION 89
S-CURVES AND INNOVATION 91
WHY INNOVATE? 92
TAKEAWAY 93

PART 2. GETTING THERE: INGREDIENTS FOR EXECUTING ON INNOVATION **94**

5. BUSINESS AGILITY AND THE DOGFIGHT IN THE CLOUD **95**

OODA Loops in the Cloud 109
Takeaway 112

6. INNOVATING INNOVATION ITSELF **114**

An Architecture for Innovation 116
Foresight and Insight 118
Frameworks: The New Role of Strategy 120
Takeaway 124

7. OPEN INNOVATION: CAST A WIDE NET **126**

The Tale of Stone Soup 126
The Gold Standard *128*
GE's Ecomagination *128*
Dell's IdeaStorm *129*
Procter & Gamble's Connect and Develop *130*
IBM's Global Innovation Jams *133*
Weak Ties and The New Polymaths 134
Some Open Innovation Issues 136
Open Innovation Platforms 138
Takeaway 142

8. INNOVATION AND SOCIAL NETWORKS **144**

Listening Posts 147
Persona Management Systems 153
Takeaway 156

9. COLLECTIVE INTELLIGENCE: THE DELIBERATORIUM AND ARGUMENTATION SYSTEMS **158**

Takeaway 163

10. BUSINESS PROCESS INNOVATION IN THE INTERCLOUD **164**

Takeaway 169

11. BUSINESS INNOVATION: CULTURE AND THE PROCESS **170**

A Quick Recap 170
The Culture of Innovation 171
The Innovation Process 172
Building Blocks for Sustainable and Systemic Innovation 177
Plan of Intent (Observe and Orient) 178
Plan of Record (Decide and Act) 182
Takeaway 186

12. INNOVATION IS A TEAM SPORT: PLAY IT TO WIN 187

Collaboration and the New IT Stack 191
Human Communication with Implicit Collaboration 193
Explicit Collaboration Via Human Interaction Management Systems 197
Takeaway 204

13. BIG DATA AND PREDICTIVE ANALYTICS 206

Takeaway 218

14. THE END OF MANAGEMENT —AS WE KNOW IT 221

Open Leadership 225
Takeaway 233

15. SERVICES INNOVATION 234

Getting There 241
Takeaway 245

16. WHO YOU GONNA CALL? THE FUTURE OF IT SERVICES 246

The Rise of the Cloud Broker 253
Whither the Internal IT shop? 256
Takeaway 258

17. THE FUTURE OF THE CIO 259

Setting the Stage for a CIO rEvolution 259
You Can't Judge a Job by Its Title 261
Today's Realities 263
Takeaway 265

18. THE FRACTAL COMPANY 266

The Current State of Business 266
Fractal Enterprise Architecture 267
Is There a Mobile Autonomous Intelligent Agent in the House? 271
The Great Dance of Business 272
Takeaway 274

19. EPILOGUE: THE AUDACITY OF INNOVATION 275

Let No Collapse Go to Waste 282

APPENDIX A: BIBLIOGRAPHY 285

INDEX 286

ABOUT THE AUTHORS 297

Foreword by Derek Miers, Forrester Research

And There Goes Another One – Who Knows? Every day, new things emerge – new capabilities that just weren't there yesterday. There goes yet another innovation that could transform the competitive landscape of your industry. And as this happens, some fade away; getting lost in the sea of competing business models.

As we go to press (literally in the last week), Amazon launches a new process support capability embedded in the heart of their cloud offering. With tremendous scale and elasticity, what does it enable that wasn't there before? Will a new ecosystem of tool and application developers emerge to leverage this coordination capability in the cloud? What will Google do? Should Microsoft join the fray? Who knows?

As we go to press, EMC demonstrates a secure virtual cube of business capabilities – processes wrapped up with documents, applications wrapped up with cases, wrapped up with change projects, wrapped up with your systems of record – specialized by country or business unit and then delivered to a public cloud, to a private cloud, on premise, or a blend of all three; delivered differently at different phases of your business model development. What does that mean for the shared services organization trying to span the globe and support 150, 000 employees? What does it mean for the corporate CIO? What does that mean for the BPO industry, or other big players? For the small players, or for start-up entrepreneurs? Who knows?

As we go to press, a small but successful tech vendor (AgilePoint) shows how thousands of distributed workers can co-create a set of documents together – collaboratively in real-time with a shared view of the content while working in familiar applications like Word, Excel and PowerPoint – tracking changes, discussing and commenting on each other's ideas. All of that delivered via Microsoft's Azure cloud offering, enabling firms large and small to leverage their hot innovation. What does this mean for the publishing industry? What does it mean for the Oil and Gas giant with 30,000 well heads that need documenting individually? How will this innovation influence an MBA class in Stanford as they work on their next joint project? What about the impact on the relationship between a legal counsel and her client on a corporate takeover? Who knows?

And all of that in just the last week! What will the next month, the next quarter or the next year bring that wasn't there before? Will any of these firms realize their vision of supporting frictionless commerce – of instantly outsourced market places (with themselves at the core); a place where disaggregated specialist providers vie to support your business model, hustling for a share of your pie? Who knows?

Will these sorts of innovations favor incumbent players? Or will they usher in a whole new era of specialist providers that outmaneuver slower moving dinosaurs? Will your organization be one of the dinosaurs, or will you seize the opportunity to reinvent the relationship you have with customers and the business model you use to deliver value?

As we go to press, Citigroup announce they will adopt IBM's Watson technology to explore the use of deep content analysis and evidence based learning capabilities to help advance customer interactions, improving and simplifying the banking experience. Now here is an example of a major corporation that has decided to do something. But what does that mean for competitors in the ultra-competitive financial services domain? How could the travel industry use this technology? What will Big Data do for your competitive landscape? Who knows?

Let's face it – computers, the internet of things and communication technology can support just about anything you might dream up; but what's mostly missing is the ability to dream up what it all means in a given situation. How can we break the habit of thinking linearly embedded in our Western culture since the time of Newton and Descartes? What opportunities exist to create parallel Business Models that leverage the inherent functionality of the Web? Rather than value chains, it's value networks and webs; it's ecosystems of partners and suppliers. If one node fails, the network reconfigures itself on the fly.

While technology innovations continue to march on – driving an exponential number of potential new business options, what's often missing are the methods and critical thinking needed to generate those options, and then sift through the possibilities. A new swathe of entrepreneurs is entering the market, rapidly testing new business model hypotheses. The question is whether your organization can develop a flexible, adaptable "target operating model" approach – think of it as a way rapidly clarifying and fleshing out strategic intent, enabling you to translate that into efficient and effective operating business.

Who knows what the implications are for you and your organization? Will you be caught in the headlights, overpowered by the sheer number of potential possibilities, perhaps dealing with the first wave of change but then getting flattened by the tsunami that follows? Only you, and what you choose to do, will affect that outcome. *Business Innovation in the Cloud* is designed to give you a set of insights and an actionable approach to help you dream; then execute on those dreams. As Stikeleather and Fingar caution us, ideas are easy, but *execution is all* when it comes to business innovation and developing new target operating models. In the end, if someone is going to eat your lunch – shouldn't it be you?

Notes

Preface

In many ways, there's nothing really new about this thing we call "innovation." Search "innovation" at Amazon and you'll get 44,000+ results. What else can be said?

And then there's this newfangled thing called "cloud computing." Search Amazon and you'll find 900+ books on the subject published since 2009.

But the search results are paltry to none when you search for books at "the intersection of business innovation and cloud computing." Actually there's a lot to say at the intersection of business innovation and cloud computing, for *execution* is all when it comes to innovation. The platform where innovation takes place has changed, and changed dramatically with the advent of global hyper-connectivity and the advent of cloud computing.

This book is laser focused on the intersection of business innovation and cloud computing. How does the Cloud influence, enable, facilitate and accelerate innovation in business, and how does innovation leverage the Cloud while expanding the reach and value of the Cloud itself?

The Cloud is enabling innovation to take place faster and faster. Cloud computing is removing a tremendous amount of economic friction from the marketplace so that companies of any size do not have to worry as much about the costs of creating, selling and distributing new products and services, or supporting those products and services. As a consequence we will see an explosion of new start ups by entrepreneurs and intrepreneurs inside large enterprises; an explosion of new products and services resulting simply because the cost of failure has been so reduced; and global market reach so easily achieved.

A new generation of *digital natives* is facile with the new technology, new work patterns, new approaches to doing work and higher levels of collaboration. Most creativity and innovation comes from collaboration, multiple minds coming together. Collaboration enabled in the Cloud makes it easy for companies to work together allowing each to work on what they do best to create value for their customers.

Okay, so much for what we know about the relationship of the Cloud to innovation. There's also what we don't know, like at the beginning of the Web. Way back then, in Internet time, we couldn't begin to imagine the new business and economic models: giants like Amazon, eBay, Facebook and Google; and totally different ways to access information and services like smart phones and tablet computers.

Although none of us can know the unknowable, we can strive to gain the agility we need to manage the unexpected, the unpredictable.

Thus while we have no magic formulas for innovating our way into the future, we can arm ourselves with the management approaches and tools needed to flow with the future. And that's not some esoteric view of innovation, it's about *executing on innovation.*

Executing on innovation isn't a one time affair, it's a systematic process that is built for change. And change is the only constant in today's business world.

This book strives to provide insights and outline strategies for executing on innovation with cloud computing, the next quantum leap in the use of information technology in business.

Our goal is to provide a starting point that will help you know what is coming, how to think about addressing it and maybe some places to get started.

We wish you great success in your journey ahead, doing the work of business innovation.

Jim Stikeleather and *Peter Fingar*
Tampa Florida, 2012

Jim's Acknowledgement

Every piece of work is a product of the works that preceded it. Much of this book draws on the work of people like Gary Hamel, Jim Champy, Daniel Pink, and too many others to mention. These are some of the people who realized that this thing we call management is not received wisdom, but a technology that can improve and evolve to address the changing environment it finds itself in.

Every person's ideas come from the ideas of others – expanded, contracted or applied in a new way. More importantly, the best ideas are those that arise from the challenges of others. I want to specifically thank my colleagues—Pamela Cawthorn, Robert Marshall, Doug Reeder, Glenn Wintrich, and of course, my coauthor—for all their assistance in developing and challenging many of the ideas that are presented here.

Everyone I have met, worked with, played with, talked with, read or simply listened to over the last five years has contributed to this book—lessons from motorcycle and car racing, pistol shooting, watch making, football, soccer, hockey, all provided insights into what is happening in the world. As Yogi Berra said, "You can see a lot just by observing." Or as Keith Code from California Superbike School said, "Get your eyes up and look broadly."

And special thanks go to my family, especially the youngest ones. It takes a lot of patience and understanding to accept that Granddaddy didn't have much time to take you to the park and push you on the swing until he finished the intense work of this book.

Part 1.
The State of Business Innovation

Before we can properly discuss the changing role of cloud computing as an enabler of business innovation, we need to set the *context* of what's happening in the larger socio-economic world *outside* of business, what's happening in our world as a result of dramatic advances brought about by the global hyper-connectivity of the Internet and the accelerating advance of technology. For example, watch how today's technology can remix Van Gogh's classic, *Starry Night* into something new:

tinyurl.com/cal2wtw

For the want of an innovation, creativity was lost.
For the want of creativity, the idea was lost.
For the want of an idea, the product was lost.
For the want of a product, the customer was lost.
For the want of a customer, the business was lost.
For the want of a business, the company was lost.
All for the want of a single innovation. —Louis Patler

1. Prologue: Something is Happening

"Something is happening,
but you don't know what it is,
do you, Mister Jones?"
—Bob Dylan, *Ballad of a Thin Man.*

"Toto, I've a feeling we're not in Kansas anymore."
—Dorothy, the 1939 film, *The Wizard of Oz*

There's an ancient Chinese proverb that says, "May you live in interesting times." While purporting to be a blessing, the proverb is a curse when the expression is used ironically, with the clear implication that "uninteresting times" of peace and tranquility, are more life-enhancing than interesting ones. Although the Chinese language origin of the phrase, if it exists, has not been found, that doesn't detract from its usefulness, especially in framing what is happening in business and in the world in general today.

Perhaps another useful insight into what's happening today could be borrowed from John Naisbitt, author of *Megatrends*, in that we are in "a time of the Parenthesis," the time between eras. "Although the time between eras is uncertain, it is a great and yeasty time, filled with opportunity. If we can learn to make uncertainty our friend, we can achieve more than in stable eras. In stable eras, everything has a name and everything knows its place, and we can leverage very little."

"But in the time of the Parenthesis, we have extraordinary leverage and influence—individually, professionally and institutionally—if we can only get a clear sense, a clear perception, and a clear vision of the road ahead."

While our times are interesting (in both cursed and blessed ways), the idea of the Parenthesis is a neutral image of an opening and space between times, an opportunity to pause (for just a second) and consider the limitless possibilities for change—that we can make either good or bad.

What is happening around us?

Everything 2.0

The term "Web 2.0" was coined in January 1999 by Darcy DiNucci, a consultant on electronic information design. As the term gained traction around 2003, it came to mean Web applications that facilitate participatory information sharing, interoperability, user-centered design, and collaboration on the World Wide Web. A Web 2.0 site allows users to interact and collaborate with each other in a social media dialogue as creators (prosumers) of user-generated content in a virtual community, in contrast to Web sites where users (consumers) are limited to the passive viewing of content that was created for them. Examples of Web 2.0 include social networking sites, blogs, wikis, video sharing sites, hosted services, and Web applications.

But now the term has led to just about "Everything 2.0" where participation and co-creation of value are moving to the core of many human activities:

- *Enterprise 2.0* centers on the way we organize work. New work structures can enable contextual, agile, integrated, and simplified collaboration among distributed workforces and networks of partners and customers. In effect the extended enterprise replaces the monolithic enterprise resulting in more, yet smaller, flatter, distributed and focused organizations. This totally disrupts traditional culture, strategy and infrastructure. We are moving from Enterprise Systems Architecture to Social Systems Architecture and "Bossless Organizations."[1] As this happens, what are the roles for traditional managers and executives?
- *Management 2.0* is being described and developed by professor Gary Hamel and others that can be seen in the work, references and activities at the MIX (Management Innovation eXchange)[2] as management tries to adapt to new models, new realities and a new workforce. Better science (e.g., behavioral, cognitive) is showing that many of our assumptions about what motivates and dissuades people are often wrong—especially when knowledge, creativity and innovation are desired. A short video by Dan Pink sums this problem up nicely, all of which suggests our current assumptions about managing are inconsistent with our current times.

Dan Pink

tinyurl.com/2ga47re

- *Capitalism 2.0* is discussed and debated around the globe. Do corporations exist solely to maximize their bottom lines? Is *value creation* very different from *profitability* as we measure it today? Almost simultaneous deterioration of major sectors of the economy suggests that something is amiss. Bailed-out auto makers lost their ability to listen to markets and make things people wanted; print media has lost its relevancy as a market maker for advertising; financial institutions created opaque, incomprehensible and unsustainable markets; and sanctioned monopolies (telecom, electric, water) have fought any attempt to open their markets, even though opening their markets and becoming better middlemen is their opportunity. Do speculators make too much and creators not enough? Is all this because Capitalism 1.0 is built on an obsolete set of ideals that will be superseded by Capitalism 2.0?

- *Economics 2.0* embodies an assertion that traditional economic behavior isn't behaving anymore, and no one is really sure why. Traditional government stimuli do not seem to be having their historical effects. Modern economists are starting to think about growth in both developed and emerging economies and contrast those ideas with earlier views in economics. The focus of the modern understanding is on ideas and the ability of ideas to improve technology, leading to prosperity rather than capital and capital assets. Unlike physical capital, ideas can be enjoyed by many people at once, explaining why past models that ignored ideas and focused on physical capital failed to account for the magnitude of economic development in places like Silicon Valley. Behavioral economists have attacked many of the assumptions of Economics 1.0, including efficient financial markets and rational expectations. Mathematicians are starting to engage using the principles and math of complex adaptive systems theory to better describe what goes on in economies.[3] Does this suggest we don't fully understand what creates value?
- *Information Technology 2.0* represents a new paradigm for delivering information, coordinating work, collaborating on ideas, and reaching customers. IT 2.0 enables all the other "2.0s." Information Technology is effectively becoming the air and water of economic activities—it's essential, pervasive and basically free. Things we managed so diligently in the past—hardware, software, data centers, even to a degree the people who create them (think open source)—are disappearing as first order considerations in business, hidden behind ideas like "the Cloud," "fabric," and any number of "as a Service" offerings. Even leading-edge technology developments divorce themselves from the physical capital aspects of IT and focus on the conceptual and idea-driven elements such as *Content-Centric Networking* ...

tinyurl.com/3qcwe9

So what is the new role of IT and can we do a better job of managing it than we have done with air and water?

TED's thecity2.org is one example of the many more "2.0s" out there—our approaches to government and sovereignty, our social norms and values, our perception of rights and obligations, our methods of education (online learning, mastery models, project-based learning), even an ever expanding plethora of life styles. All these are creating new demands, wants, and opportunities for business just as we are discovering that our old business heuristics don't work anymore.

So what is a business leader supposed to do? Well, the good news is we have gone through these parenthetical times many times and in many areas of human life. We always get through them, and the best way is by understanding how we got through them before.

When Memories Exceed Dreams, the End is Near

As the coauthor of *Reengineering the Corporation*, the late Michael Hammer wrote, "The hallmark of a truly successful organization is the willingness to abandon what made it successful and start fresh."

The transition (upheaval?) that we are about to go through in our paradigms of economics, business and management is similar to that the science of physics passed through early in the 20th century. A good summary of that transition can be found in the footnote.[4]

One could argue that we are entering the quantum mechanics phase of management—where our "natural" (Newtonian Classical Mechanics) instincts are wrong about what is really happening (Quantum Mechanics and General Relativity).

On one hand, this book is a first order approximation on what is happening and what seems to work. Its goal is to provide some jumping off points on how to address all the issues and opportunities that are coming at contemporary executives, and some references to other ideas and thinkers to help understand what we are encountering.

On the other hand, the book assumes that IT will be the enabler, facilitator and accelerator for all that is happening, but not necessarily the cause, and definitely not a tool to inhibit or prevent the coming "2.0s." Hopefully we provide some insights as to how to move your IT to the next level. Historically (in practice if not in philosophy), IT has suffered from "Marketing Myopia," a term used in marketing as well as the title of an important paper written by Theodore Levitt in 1960 in the *Harvard Business Review.* Marketing myopia refers to "focusing on products rather than the customer." George Steiner claimed that if a buggy whip manufacturer in 1910 defined its business as the "transportation starter business," they might have been able to make the creative leap necessary to move into the automobile business when technological change demanded it.

It is our contention that IT should be in the innovation business (enabling, facilitating and accelerating ideas and their implementations) and that "the Cloud" is the "air and water" needed to make it happen.

As we observe what is happening, we see that value creation is less a linear progression of applying capital, thought and process, but more emergent, unpredictable and more driven by people coming together to solve problems as they appear.

Ideas, conversations and outcomes become more important than process, productivity or traditional outputs (products and services) of a business.

Value is less a product of the inputs (raw materials, time, IP, labor) than the resulting value of those outputs as determined by, not the market, but by each individual consumer of those outputs and the outcomes they achieve.

Value, and the resulting price, becomes situational—Who are the consumers? Under what situation and environment are they consuming? With whom they are consuming? Why are they consuming?—with each consumer's consumption generating different outcomes each time.

Traditional management goals of productivity are divorced from delivering value in this new world where "doing the right things is more important than doing things right."

This suggests lots of trials and lots of failures—so do them quickly and learn.

All this tells us a couple of things.

First, *the* key to success is now *innovation*, not cost controls, distribution channels or messaging, though they are all still important.

Second, information technology is the key to enabling, facilitating and accelerating innovation. Traditionally, technology was applied to reduce human labor to increase efficiency and reduce costs, but with value arising out of consumptive contexts, technology must shift to enabling, facilitating and accelerating communications, collaboration and serendipity among all of the players (suppliers, consumers, partners, regulators, and planners) and potential contributors of value.

Information technology's role going forward, as the platform for supporting innovation, is to increase the breadth of potential collaboration (number and variety) and the depth (level of detail and interaction) of collaboration. The goal should be to reduce or eliminate economic frictions (time, distance, total costs of interaction, learning curves, risk, initiation, and so on) as well as reduce or eliminate transaction costs (What does it take to get distinct parties to interact and serve each other, particularly in relation to time and capital?).

We believe that the Cloud will be the major enabler, facilitator and accelerator for innovation in any and all industries, markets and many if not most human endeavors—not as a technology but as a totally new paradigm for organizing, managing, and executing business processes. We believe that innovation will be the one true form of competitive advantage when the Cloud eliminates many of the scale, cost and reach barriers incumbent companies have built over the years. These are some of the reasons we wrote this book.

Further, it can't all be random and happenstance. Innovation, while not linearly, moment-by-moment plan-able or manageable, can be organized and cultivated.

We hope you will get some ideas from us on how to execute on this thing called *innovation* in the 21st century.

Ideas are easy.
Execution is all,
when it comes to business innovation.

What if? ...

Three Laws of Prediction:

- When a distinguished but elderly scientist (technology expert / business consultant) states that something is possible, he is almost certainly right. When he states that something is impossible, he is very probably wrong.
- The only way of discovering the limits of the possible is to venture a little way past them into the impossible.
- Any sufficiently advanced technology is indistinguishable from magic.
  —Arthur C. Clarke, 2001 Space Odyssey

"Modern economies are not built with capital or labor as much as by ideas. The rise of the Cloud is more than just another platform shift that gets geeks excited. It will undoubtedly transform the IT industry, but it will also profoundly change the way people work and companies operate." —*The Economist*, "Let it Rise."

What if your company *could* ...

- Follow the sun and do business with customers around the globe 24/7?
- Use the world's largest supercomputer and pay for only as much or as little as you want without having to buy it, house it or provision it?
- Gain business insights derived from deep analytics that scoured the world's Big Data (social networks, videos, emails)?
- Connect and collaborate with your customers to co-develop the exact products and services they demand?
- Connect your business processes with suppliers and business partners around the globe to build unique new value propositions for your customers?
- Rapidly pilot and test out new products and services, then scale up those that fly, and shut down those that don't?

Well, just rearrange the letters in the word "could" and you have the word "Cloud." In a stroke, "what if you could?" becomes "you *can*." You *can* make profits out of thin air.

To paraphrase what the legendary computer scientist Alan Kay said about the Dynabook back in 1972, "If the mind can conceive it, the Cloud can Achieve it."

What's the Dynabook? It's the concept of personal computing that has provided the foundation for commercial products such as the Macintosh and the iPad and other tablet PCs. Here's an update from Kay regarding the iPad along with his 1972 paper, "A Personal Computer for Children of All Ages."

Alan Kay and the Dynabook

tiny.cc/vkusn

tiny.cc/acza1

Although we are tempted by the press to think that certain people invented the tablet computer or the Internet, the concepts were in fact "invented" decades ago by real scientists, not those who commercialized the inventions. Soon we may hear outlandish claims of who invented the Cloud! Don't believe them, for there's no such "thing" as the Cloud or cloud computing. If anything, it's what the Internet was intended to be when designed in the 1960s: an endless network of computers made up of networks of computers.

In 1961, in a speech given to celebrate MIT's centennial, John McCarthy was the first to publicly suggest that computer time-sharing technology might lead to a future in which computing power and even specific applications could be sold through the utility business model, like water or electricity. Enter today's Cloud.

The Cloud is not a new technology. It's not a new methodology. It's not a new architecture, though it leverages better implementations of the technologies, methodologies and architectures of the past.

So what *is* the Cloud?

The Cloud is a new economic delivery model that enables new business models and processes supporting rapid-fire business innovation.

The Cloud is a business game-changer for companies large and small. It matters not if your company is you, a sole proprietor, or a huge multinational corporation.

The playing field is leveled in the Cloud—one shared world, one shared computer, one shared information base.

Rapid elasticity is at the heart of cloud computing: use as many or as few resources as you need—scale up or scale down on the fly and pay as you go. By providing flexibility and adaptability, the Cloud will also accelerate the introduction and adoption of new technologies, methods and processes; speeding innovation and business evolution across all sizes of business and industries, simultaneously.

Think the computers in the world of Star Trek—and before you laugh, think about the communicator attached to your belt (smart phone). In general science fiction tends to underestimate future capability rather than overestimate it. Some of what we talk about in this book is for effect, to get you to think—but keep Amara's law in mind as you read this book, "We tend to overestimate the effect of a technology in the short run and underestimate the effect in the long run."

The Cloud is an enabler *and* a driver of business innovation that allows you to:

- Rapidly co-develop new products and services with customers, suppliers and business partners.
- Collaborate at the process level to implement new multi-company, end-to-end value delivery systems.
- Achieve rapid time to market with pay-as-you-go resources versus making capital expenditures.
- Form and later disband Virtual Enterprise Networks that allow small companies to band together to compete with or collaborate with large companies for a given market opportunity. Little fish need big fish operating resources and big fish need the innovation of little fish in today's marketplace. Keep in mind that the Cloud is creating an even more expansive ecosystem and Gause's law, a proposition that states that two species competing for the same resources cannot coexist if other ecological factors are constant. When one species has even the slightest advantage or edge over another, then the one with the advantage will dominate in the long term. In short, Gause's competitive exclusion principle means "Complete competitors cannot coexist," which also means that with the Cloud's ever expanding ecosystem, we all need to learn, or re-learn, *coopetition* (cooperative competition). Principles of competitive structures have been described in game theory, a scientific field that received great attention with the book, *The Theory of*

Games and Economic Behavior in 1944.

- Fail fast and furiously, leveraging Gall's law. Gall's Law is a rule of thumb from John Gall's *Systemantics: How Systems Really Work and How They Fail*, "A complex system that works is invariably found to have evolved from a simple system that worked:"
    - Beta test new products and service in a sandbox before making large scale implementations.
    - Rapidly scale up new products and services that pass the beta tests.
    - Instantly shut down products and services that don't pass the beta tests
- Connect and co-develop innovative products and services with market leaders through their open innovation offerings.
- Scale up or scale down resources needed in response to customer demand and resource usage.
- Tap Web-scale, Big Data analytics to gain market insights from massive unstructured information pouring thru blogs, wikis and social networks.

> Just because you can do something doesn't mean you will do that something. The "doing" takes work and most contemporary organizations aren't designed for doing the work of innovation—they are designed for doing what they already do that got them to where they are.

Companies, and IT groups within them, continuously strive to improve current productivity, but breakthrough and disruptive business innovation means *not* doing what they already do.

When they do launch an innovation initiative, cognitive dissonance, that uncomfortable feeling caused by holding conflicting ideas simultaneously, sets in. They devour countless books and articles in search of support for unleashing the creativity that they believe will lead to breakout ideas. They search for "innovation" at Amazon and, as we've mentioned get pointed to over 44,000 titles.

Many overlook what Thomas Edison said over 100 years ago, *"Genius is 1% inspiration and 99% perspiration."*

That other 99% represents a long and trying journey. This book takes on that journey as its central theme. The difficult journey from New York to San Francisco that took early pioneers years with horse-drawn wagons now takes but a few hours as a result of the modern airliner. And keep in mind, that those earlier pioneers weren't 100% sure where they were going when they started (an anathema to current goal-driven management practices), they were just heading west. And they weren't sure of what specific route they were going to use (an anathema to contemporary business planning) but adjusted the route based upon progress, time, season, weather, hostiles, friendlies, and whatever else they encountered as they progressed.

Today, the business innovation journey can be transformed in a like manner with the advent of cloud computing. Because we now live in a Knowledge Age, the Cloud can change the rules and the pace of the innovation game. To leverage the Cloud, you only need to know the general direction you are heading, like the early pioneers did.

> In the past, IT was about productivity. In the Cloud it's about collaboration, a shared information base and collective intelligence (the wisdom of crowds, social networking and crowd sourcing).

In the past business concerns centered on owning the materials and means of production. In today's global economy with low-cost commodity suppliers scattered around the globe, ownership of production assets is displaced by multi-company supply chains that stretch from Shanghai to San Francisco. Even knowledge itself is now a global commodity.

From the Industrial Age to the Knowledge Age and now on to the Innovation Age, all is changed at each major step along the way in economic history. In the Innovation Economy, smart companies will choose to do the work of business innovation in the Cloud just as they would choose an airliner to travel from coast to coast instead of horses and covered wagons.

Enterprises express a desire to be innovative. It is the oft-cited statement from countless corporate leaders who evoke it in mission statements and pay homage to its value in supporting corporate growth. But, like many corporate goals, innovation is often more a vague, conceptual symbol than a clearly defined process. On closer examination, it is obvious that few organizations truly commit to innovation in a measurable way with head count, process management, funding or a meaningful reporting relationship to the company's executive leadership. Or worse yet, organizations set up a laundry list of tactics as a basis of an innovation strategy (we will buy this tool, we will train these people, and so on.)

For most of the past decade, economic growth was booming and a precise definition and understanding of innovation may not have been an imperative. The Great Recession of 2007+ brought an abrupt end to the boom times. Now competition is being reshaped and the pace of business is faster and more volatile. Such turbulence makes the need to understand and harness the innovation *process* far more important than ever before.

The ability to innovate may well be the single most important capability that companies have in helping them sustain a competitive advantage. The need to innovate applies to processes, technologies and organizational structures, as well as management and leadership styles. Those must be innovated first, otherwise there will be no innovative products and services (note that in the future, there will be no such thing as a product without services embedded, and the Cloud will be the most efficient and effective way to deliver services). Organizations that want to develop innovation as a core competency must create a culture in which innovation can thrive, and then extend the innovation process to the entire employee base. They also need to set clear objectives and allocate both budget and personnel resources to establish enterprise innovation as the de facto approach for achieving business growth and operational excellence.

But beware of Goodhart's law, "When a measure becomes a target, it ceases to be a good measure." Too often goals become focused on the measures (rather than the outcomes) and the desired outcomes become lost. The outcomes of an innovation process are innovations, and they are determined solely by the market, not internal measures. Measures, like maps, are good for navigating—where are we, how far have we come, how far do we have to go—but remember that the map is not the land.

Questioning Innovation

A dictionary definition of innovation is "introducing something new." But these days, with the rise of the buzzword "innovation" in the business literature, you might be led to think that innovation itself is something new. Let's see if we can net out what's new about innovation by asking a few questions about this hot business topic.

What's New About New? Here we go again with yet another hype curve, the Innovation Hype Curve. Why all the newfound interest in innovation? In short, it's globalization. Executives fear their companies becoming commoditized as a result of total global competition, and they desperately seek new ways of distinguishing their products and services so they can continue to earn healthy margins. But even though companies know they must innovate, it's not clear exactly how to go about it in the complex global markets of the 21st century. Is there more to it than just inventing some new gizmo or thingamajig?

And it is not just globalization in a geographic sense, but also in terms of how others can compete with you. The Cloud—the *technology* Cloud, the *business services* Cloud, the *information* Cloud—as it evolves eliminates so many economic frictions (costs, capital, resources, scalability, availability, reliability, survivability, time, etc.). Even the smallest least-funded organization anywhere on the globe can *appear and deliver* like the largest. That means less money to "try an idea." And that means more investors are able to create more innovations that can operate in smaller and smaller sized markets—challenging the 20th century management assumptions about scale benefits.

Isn't Innovation Just Invention? 3M's Geoffrey Nickelson once described the critical point about invention versus innovation, "Research is the transformation of money into knowledge. Innovation is the transformation of knowledge into money." It was Dan Bricklin and Bob Frankston who co-invented the spreadsheet (Visicalc) that launched the microcomputer as a mainstream business tool. Of course, it was Bill Gates who ultimately transformed this digital invention into real money—and that's "business innovation." So, one key aspect of innovation that's new is the changing focus from invention to innovative new business models and initiatives that bring compelling new value to the marketplace.

Isn't Innovation About R&D Labs? Secretive R&D labs used to be the main source of invention. In 1900, Charles Proteus Steinmetz convinced the GE leadership that they would need a research laboratory, and the first industrial research lab in the U.S. was born in the carriage barn in Steinmetz's backyard. Today, GE Global Research consists of 2,500 employees working in New York (a few miles from the original barn), Bangalore, India, Shanghai, China, and Munich, Germany. But there's even more to this story; it's called Open Innovation. Open Innovation changes the game of where ideas come from; they are no longer the exclusive property of internal R&D, marketing, and product development organizations.

What's Open Innovation? The new world of widely distributed knowledge has led to the business proposition of "open innovation." No longer can companies win the innovation arms race from the inside-out (internal R&D). They should, instead, buy or license innovations (e.g., patents, processes, inventions, etc.) from external knowledge sources, turning the table to outside-in innovation. In turn, internal inventions should be considered for taking outside the company through licensing, joint ventures, spin-offs, and the like. Procter & Gamble, the poster child for open innovation, is determined to move to the position where half of its ongoing stream of innovation comes from outside the company. Don't confuse open innovation with open source, free software. Open innovation is all about the money to be made. In short, in today's wired world, *knowledge can be transferred so easily that it seems impossible for companies to stop it.* Instead, smart companies are going to the ends of the Earth in search of knowledge that they can transform into money.

Aren't Customers the Target of Innovation? No, now they are the source! The Industrial Age was about mass production. Innovation was R&D-driven, from the inside-out. It was about supply-push. The Customer Age is about mass customization. Innovation must now be driven from the outside-in. It's now about demand-pull. It is about turning a company, and its entire value chain, over to the command and control of customers.

This new reality demands a shift in our thinking about innovation. Winning companies will be so close to their customers, they will be able to anticipate their needs, even before their customers do, and then turn to open innovation to find compelling value to meet those needs. That, in turn, as pointed out by the late management guru, Peter Drucker, *means becoming a buyer for your cus-*

tomers. That means "business mashups" where your company joins forces with suppliers, sometimes even your competitors, to expand your product and services offerings, blurring industry boundaries. Because *your customers are your only true asset in the new world of low-cost suppliers,* your business model will likely need to be expanded so that you can fulfill as many of your customers' needs as possible. This is how smart companies are seeking true growth and avoiding commoditization—they are no longer in the product business; they are in the customer business.

Isn't Innovation An Episodic Event? The light bulb. The steam engine. The airplane. Innovation isn't just about some star in the R&D suite delivering a radical breakthrough, a home run, every now and then. What's wrong with hitting singles and doubles with regularity? Isn't it time that we stopped just talking about out-of-the-park innovation and got serious about developing the capability required to manage the complete innovation lifecycle? And companies better manage that lifecycle because in today's wired world competitors can catch up to your innovation in an instant. Just look at the glitzy new Apple iPhone. Hackers broke the code requiring the exclusive use of AT&T, a Chinese company has a clone under development, and Google is pursuing the gPhone. In just weeks after its introduction, Apple had to cut $200 off of its price due to competitive pressures. Today, it's not a single innovation; *the challenge is to set the Pace of Innovation*—once you innovate, then you really have just begun and will need to run hard to outpace your competition.

Isn't Innovation Just About Great Ideas? There is certainly no lack of great ideas in today's hyper-connected, wired world. But companies report a 96% failure rate on their innovation attempts. What's up with that? It's largely a result of innovation being approached in a haphazard fashion. Today, however, leading companies are taking the haphazard approach to business innovation and turning it into systematic, repeatable, business processes (think Innovation Process Management or IPM). IPM can be compared to the rise of the total quality movement in the 1980s, where leaders such as Toyota taught the lesson of quality-or-else. It turns out that quality is all about process.

Ditto for innovation.

Some companies have already implemented systematic approaches to innovation process management. GE calls it CENCOR (calibrate, explore, create, organize, and realize). The Mayo Clinic calls it SPARC (see, plan, act, refine, communicate).

A caution: Some companies have almost killed their ability to be innovative in pursuit of efficiency in their *current* business processes. 3M is an example. At the company that has always prided itself on drawing at least one-third of sales from products released in the past five years, today that fraction has slipped to only one-quarter.

Those results are not coincidental. Efficiency programs such as Six Sigma are designed to identify problems in work processes—and then use rigorous measurement to reduce variation and eliminate defects. When these types of initiatives become ingrained in a company's culture, as they did at 3M, creativity can easily get squelched. After all, a breakthrough innovation is something that challenges existing procedures and norms.[5]

Takeaway

It may seem like we wrote this introduction to suggest to you why you might want to read the book. But that is the nature of what is happening – so many things happening at once that feed upon and reinforce each other. A focused, in depth exploration of any one of them would miss the point and importance of the sum of all of them. The good news is that disruption, risk and struggling are all things that happen as we go through change, and dealing with these challenges build strength, character and long-term competitive advantage for companies. The bad news is we are entering a time when it all seems to be happening at once.

We've touched on some key questions companies should ask as they position themselves for total global competition in the 21st century. Forget the old idea that innovation simply means product invention. We've seen the need for systematic innovation processes. We've seen that innovation certainly isn't new. Innovation has been around since the harnessing of fire by early man.

A new generation of innovation in the material world began when Isaac Newton watched the apple fall from the tree. Yet another new generation of business innovation began when economist Joseph Schumpeter observed capitalism's power of "creative destruction."

Management may be the greatest invention of the past one hundred years. However, so few have learned management as a science but instead have learned it as a religion—a collection of given wisdom without understanding the *why*, *what, where* and *when* from which the ideas came. In order for the us to be successful in reinventing management to facilitate innovation, managers must learn that the why and what of their past may not work in the future.

The current generation of innovation is customer-driven process innovation, the kind that can transform business models and strategies in the new world of total global competition—the kind of process innovation that can transform innovation itself, the kind of innovation that touches, and is driven by, your customers. It's about new ways of entering new market channels, creating new value-adding services and new ways of anticipating unarticulated customer demands. This is precisely where business process management meets innovation; for you simply can't do these things without process innovation that enables collaboration across the globe. We'll explore more about process innovation later in the book.

We hope to leave you with some insight we got while reading "The Perils of Bad Strategy," Richard Rumelt's *McKinsey Quarterly* article (June 2011). According to Rumelt, "A good strategy does more than urge us forward toward a goal or vision; it honestly acknowledges the challenges we face and provides an approach to overcoming them." When you finish this book, we hope that we will have helped you develop:

- *A diagnosis:* a simple explanation of the nature of the challenge of becoming innovative.
- *A guiding policy:* an overall approach chosen to cope with or overcome obstacles to innovation .
- *Coherent actions:* steps coordinated with one another to support the accomplishment of enabling,

facilitating and accelerating innovation within your organization.

The goal of this book is to enable you to "Change before you have to," as said by Jack Welch, former CEO of General Electric.

References

---

[1] http://ceomag.in/the-bossless-organization-from-bosses-to-mentor-investors

[2] http://www.managementexchange.com

[3] http://www.santafe.edu/news/item/inet-agent-based-modeling

[4] http://en.wikipedia.org/wiki/History_of_physics

[5] www.businessweek.com/magazine/content/07_24/b4038406.htm

2. Do the Work

Do the Work

Could you be getting in your way of producing great work? Have you started a project but never finished? Would you like to do work that matters, but don't know where to start? The answer is *Do the Work*, a manifesto by Steven Pressfield, that will show you that it's not about *better ideas,* it's about *actually doing the work.*

Do the Work is a weapon against resistance—a tool that will help you take action and successfully ship projects out the door. According to author, Steven Pressfield, "There is an enemy. There is an intelligent, active force working against us. Step one is to recognize this. This recognition alone is enormously powerful. It saved my life, and it will save yours."

Although Pressfield's manifesto was written for individuals it also applies to businesses facing the pressing need to innovate. In many businesses, the enemy is *inertia*, doing what they've been doing, for what they've been doing is how they've been successful so far.

Oops.

Shift happens.

Perhaps you've lifted your eyes up from your desk to look at the larger world, the larger world outside the business world.

The world has changed. There are indeed some fierce new competitors on the block. We now have three billion new capitalists in the global economy. China, Russia and India are not wanting to compete on just labor costs, they want to compete on innovation, too.

We'll talk more about this in the next chapter, but for a preview, watch *Shift Happens.*

mkpress.com/ShiftHappens

The terms "globalization" and "commoditization" strike fear in the hearts of today's executives. In response, "innovation" is being bandied about as the new Holy Grail in business. Yet, while innovative ideas are easy, doing the work is hard.

Surrounded on the one side by never-satisfied customers, and on the other side by three billion new competitors from former socialist countries, Bruce Nussbaum has some advice for business leaders that he gave way back in *Business Week's* August 1, 2005 issue. "Listen closely. There's a new conversation under way across America that may well change your future. If you work for Procter & Gamble or General Electric, you already know what's going on."

The globalization of white-collar work is the new trend beyond sending blue-collar jobs to Asia and beyond. Even much of the so-called "knowledge economy," once thought of as the last bastion of America's economic might, has been digitized and beamed to China, India, Russia and beyond. In short, knowledge work is being digitized, globalized and commoditized. So, what's left for companies wanting to avoid commodity purgatory?

Welcome to the *Innovation Economy.*

Nussbaum describes an innovation economy: "The new forms of innovation driving it forward are based on an intimate understanding of consumer culture—the ability to determine what people want even before they can articulate it. Working in what is still the largest consumer market in the world gives U.S. companies a huge edge. So does being able to think outside the box—something Americans still do better than most. But Toyota Motor Corp. has a feel for U.S. consumers, and the Samsung Group can be pretty creative, too. Competition will surely be intense."

"You're thinking 'this is all hype,' aren't you? Just another 'newest and biggest' fad, right? Wrong. Ask the 940 senior executives from around the world who said in a recent Boston Consulting Group Inc. survey that increasing top-line revenues through innovation has become essential to success in their industry."

Nussbaum also reports that nearly 96% of all innovation attempts fail to beat targets for return on investment, leading to talk of "innovation frustration" in the corner offices.

Indeed, is it time to innovate innovation itself—as a systematic, repeatable business process.

What is a business process? It's how work gets done, which leads us back to *Do the Work,* that is, do the *work* of innovation.

That, in turn, leads to another question, "How does work *work* in a hyper-connected world of total global competition?" Not that long ago, the world of business transformed from the Industrial Age to a so-called Information Age where *knowledge* was king, and where components of Industrial Age competitive advantage (land, labor and capital) became undifferentiating commodities

Oops. Today knowledge is a global commodity, as is the ability to process and use it. Proprietary knowledge is fleeting. It is stolen via hackers. It walks out the door with employees and ex-employees. It is innocently shared in social media, and the process of protecting it (the patent process) exposes it. Often times innovation is driven by the simple knowledge that something can be done even if the exact details of *how* are not known.

In today's wired world, business success absolutely depends on innovation, not just *commoditized knowledge.*

And it's not just a single innovation, it's about setting the *pace of innovation* in your industry.

It's also about *getting into new industries as* you seek to serve the *total needs* of your customers.

In today's wired world, business innovation goes hand in glove with IT innovation. That's precisely where cloud computing comes in—*doing innovation's work!*

The World Wide Web (WWW)
has morphed into the
World Wide Computer (WWC)—"The One."

Without getting technical, cloud computing provides
organizations and individuals alike
with one shared computer, one shared information base, and collective intelligence
(the wisdom of crowds, social networking and cloud sourcing).

In short, the Cloud is the stage for *one shared world.*

We don't suggest that there will be one Cloud computer, but rather the Intercloud made up of multiple, interoperating Cloud resources and services—just as there is not a single Internet computer, but a network of networks of computers. The term "The One," coined by Kevin Kelly, refers to all users touching "The One," the Cloud.

Well, so what? The "so what" is:

- The Cloud is where organizations, people and ideas come together to do the work of business innovation.
- The Cloud is the *workspace for innovation*, the place where the work of innovation gets done.
- Those aren't raindrops in the Cloud, they're *perspiration* from high-performance teams doing the work of business innovation.
- To paraphrase a recent politician, "It's the work, stupid." And the work of innovation—across the entire "idea chain" to the entire "value chain"—is taking place in the Cloud.
- It's now time to *do the work,* the hard work, of innovation. We'll examine the ingredients for doing the work of innovation in Part 2 of the book.

In business, while employees physically do the work, it is management that sets the stage and arranges the conditions for the work to be done. That's the *work of management*, and, as we are learning, it's not easy to do the work of management in today's environment.

While the Cloud will be an enabler of innovation, it is not the starting point, instead it is management itself. Management has always been a function of the times it operated in and the outcomes expected of it. Historically you can think about it this way:

1. Classical—organizing work and getting unskilled workers to perform repetitively and consistently
    a. F.W. Taylor – scientific management
    b. Henri Fayol / Max Weber – administrative / bureaucratic
2. Delegation + Committee -organizing decision making
    a. Sloan @ GM
3. Neo –classical (behavioral / human) – motivating and developing workers
    a. Mayo / "Hawthorne" effects / human relations (rise of HR departments)
    i. Theories X, Y, Z …
    b. David Packard/Bill Hewitt – social management (wandering)
    c. James G. March
4. Management Science / operations research / quality / empirical decision making
    a. Deming
5. Modern Management – Drucker
6. Process (re)engineering (Hammer and Champy)
7. Business Process Management (Smith and Fingar)
8. Constraint management (Goldratt)
9. Contingency / situational (Paul R. Lawrence)
10. Systems theory (Ackoff / Herbert Simon)
11. Chaos / complexity theory based (biological models) / self organization / institutional learning
    a. Dooley, Senge

Likewise, innovation is a form of work and should be an outcome of management. Regardless of the time, innovation shouldn't be so hard and although we often ascribe successful innovation to a person, the truth is we should be looking more closely at the *process*, a management process. In all the previous management models and theories, a few things always hold true:

- The desired outcomes are: –Creation of customers, –Lowering of transaction costs (time, capital, resources) that do not directly create value for the created customers, –Overcoming economic friction that hinders delivering value to those created customers.
- The processes used are planning, directing, staffing and controlling (in non-prejudicial uses of those terms)
- The fundamental principles used (implementations varied)
  –Division of labor (Specialization), –Unity of command (Only one boss), –Scalar chain of command (Hierarchy of Authority), –Span of control (Number of subordinates supervised)
- And the roles played by management were always the same (though balance changed): –Figurehead, –Leader, –Liason, –Monitor, –Disseminator, –Spokeperson, -Entrepreneur, –Resources allocator, Negotiator

No matter how formal or informal the innovation initiative, using relevant management processes, principles and roles is key. Innovation management is about tackling unknown problems or unknown solutions. It's about a process for reducing innovation risk, with the goal to discover what to build before you waste your time and money. In contrast, more traditional management deals with known problems and known solutions.

There are vast numbers of innovation processes out there, but all encompass the following steps:

1. *Understand and scope* opportunities to identify your most critical assumptions, usually around what problem you are trying to solve.
2. *Capture ideas* to develop the minimum product and service concepts to deliver value (desired, sought, identified, anticipated outcome) that will allow you to learn about those assumptions and deliver the needed outcomes.
3. *Evaluate and select* by testing ideas in the ecosystem of partners, suppliers, and customers to find out if what you believe is shared.
4. *Develop and experiment* as quickly and efficiently as possible until you discover the right answers to delivering the outcomes.
5. *Implement* quickly and then scale iteratively, openly, and collaboratively to the right product and market fit.
6. *Champion* innovative products and services in order to minimize the time the ecosystem has to climb the learning curve to understand the benefit. This isn't about *sales;* it includes the internal organization, suppliers, partners, and customers—the entire ecosystem.

Keep in mind that this isn't a serial process, but is a collection of recursive feedback loops—we'll talk about essential OODA loops in Chapter 5. A serial planning mindset can increase your chances of failure because you waste time executing without leveraging your continuous learning.

Each of these steps can be opened up to external contributors, with varying degrees of difficulty, risk and reward. Current innovation research focuses on how and what to open in each step and it includes the benefits and drawbacks of each.

This book isn't meant to be a treatise on business innovation. Many pundits, gurus and scholars have already done that (see Suggested Readings in the Appendix). This book is, instead, a top-view exploration of the current state of business innovation, which then moves on to *execution*, for execution is *all* in the world of business innovation.

Execution encapsulates the idea that when you are dealing with an unknown—like how to address a new market need or create a disruptive business outcome—in contrast to a known such as how can I do what my competitor is doing, only faster, better cheaper, you have to test your assumptions in as low-cost a manner as possible, and then iterate when you inevitably find out that you were wrong (fail fast).

The difference really comes down to the difference between a *planning-based model* (think about it) of execution and a *learning-based model* (do it and see what happens) of execution where you employ all the tactics you can to learn as honestly and flexibly as you can about the unknown. It sounds simple but it can be difficult to break free of the old patterns of management in practice.

Takeaway

Great ideas are a dime a dozen.
Execution is *all* in the world of business innovation.
The Cloud is where the work of innovation gets done.
Do the work!

3. Change Itself is Changing

Listen to what Dr. Albert Bartlett has to say in
"The most important video you'll ever see." tinyurl.com/7dbev9n

"It is not necessary to change, survival is not necessary."
—Edwards Deming,

A hundred years ago, change was gradual, rising steadily, the kind of change we humans like. Today, change itself has changed, for it's no longer gradual and linear, it's *exponential* and we have no prior framework with which to deal with it. When gradual change hits the *knee* of an exponential curve, it takes off like a rocket! Just consider:

- CO2 Emissions
- The number of Internet Connections
- The Information Explosion
- The number of Mobile Devices
- The number of Genes Sequenced
- Nanotechnology
- Biotechnology
- Solar technology
- The Internet of Things
- Human Population
- The list goes on

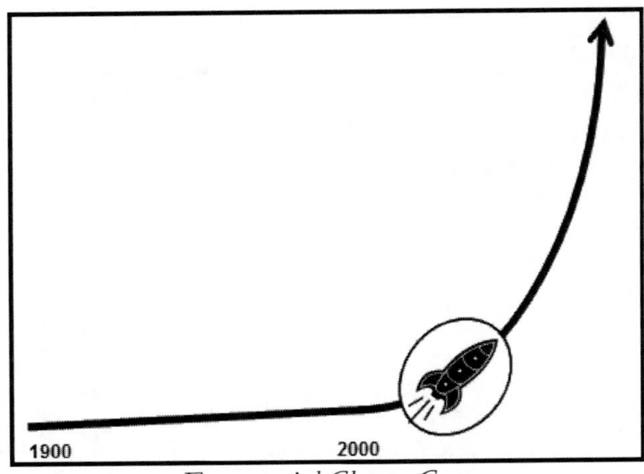

Exponential Change Curve

While we're not attempting to write a world history on change, we do want to focus on the essentials of how change, especially technological change, is affecting our business world. We become so focused on our day-to-day activities that we often don't have time to look up at the huge changes taking place in the bigger world. But it's that bigger world that will dictate what it means to be a business in the future.

Let's just look at one small area of change that will likely disrupt business models, processes, organizations, management practice, investment considerations – and pretty much everything else associated with running a business – that will require tremendous innovation that you might not be thinking about right now. Consider the data from the Global Footprint Network, "Today humanity uses the equivalent of 1.5 planets to provide the resources we use and absorb our waste. This means it now takes the Earth one year and six months to regenerate what we use in a year. Moderate UN scenarios suggest that if current population and consumption trends continue, by the 2030s, we will need the equivalent of two Earths to support us. And of course, we only have one."

"Turning resources into waste faster than waste can be turned back into resources puts us in global ecological overshoot, depleting the resources on which human life and biodiversity depend."

"If you cut down more trees than you grow, you run out of trees. If you put additional nitrogen into a water system, you change the type and quantity of life that water can support. If you thicken the Earth's CO2 blanket, the Earth gets warmer. If you do all these and many more things at once, you change the way the whole system of planet Earth behaves, with social, economic, and life support impacts. This is not speculation; this is high school science."

"Bringing an end to overshoot means investing in technology and infrastructure that will allow us to operate in a resource-constrained world. It means taking individual action, and creating the public demand for businesses and policy makers to participate."

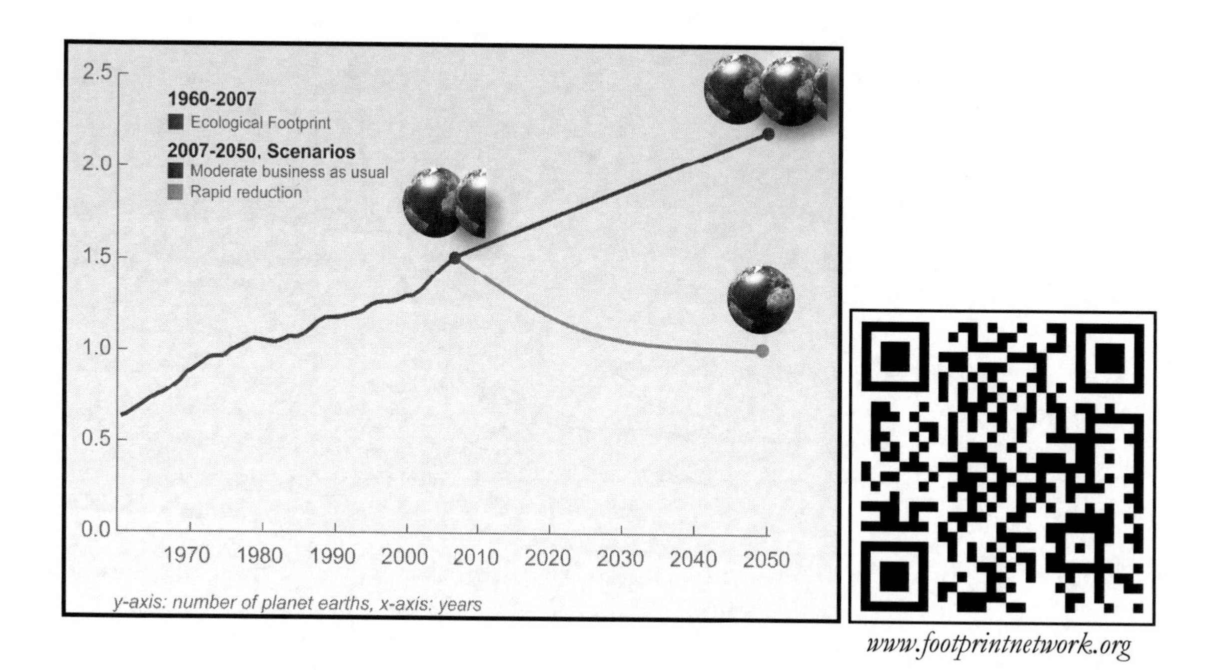

www.footprintnetwork.org

Indeed, we live in a world of exponential change. We will not change systems, though, without a crisis. But don't worry, we're getting there—today, all manner of crises loom.

Paul Gilding is an Australian environmentalist who described the coming crisis moment in his book, *The Great Disruption: Why the Climate Crisis Will Bring On the End of Shopping and the Birth of a New World,* "When you are surrounded by something so big that requires you to change everything about the way you think and see the world, then denial is the natural response. But the longer we wait, the bigger the response required." But Gilding is actually an eco-optimist. "As the impact of the imminent Great Disruption hits us," he says, "our response will be proportionally dramatic, mobilizing as we do in war. We will change at a scale and speed we can barely imagine today, completely transforming our economy, including our energy and transport industries, in just a few short decades."

We will realize, he predicts, that the consumer-driven growth model is broken and we have to move to a more happiness-driven growth model, based on people working less and owning less. "How many people," Gilding asks, "lie on their death bed and say, 'I wish I had worked harder or built more shareholder value,' and how many say, 'I wish I had gone to more ballgames, read more books to my kids, taken more walks?' To do that, you need a growth model based on giving people more time to enjoy life, but with less stuff."

Sounds utopian? Gilding insists he is a realist. "We are heading for a crisis-driven choice," he says. "We either allow collapse to overtake us or develop a new sustainable economic model. We will choose the latter. We may be slow, but we're not stupid."

Tom Koulopoulos of the Delphi Group quipped, "The most profound impact of the 20th century enterprise was in the way we moved workers to where the work was. The most profound impact of the 21st century enterprise will be in the way we move work to where the workers are."

In 2009, three quarters of companies responding to the Carbon Disclosure Project (CDP) from the Global 500 reported regulatory risk to their business related to climate change, and the same proportion reported physical risk. Climate change regulation continues to emerge in different jurisdictions and is certainly a factor in the growth of corporate climate reporting. However other drivers are also important, including pressure from investors, customers and the wider public.

Furthermore, international companies are in the unique position of having immediate and direct exposure to changing climatic conditions, and resulting changes to market or operating conditions, all over the world. A new corporate operating environment is being created that internalizes the carbon externality and makes climate change management a normal part of doing business. Today we are only part of the way there, but the rate of change is dizzying.

Just this one area of rapid change (Who would have thought managing carbon footprint would be almost as important as managing fixed assets a half dozen years ago?) has all the earmarks of disrupting our organizations (bring work to the workers versus workers to the work). The management reality is that as the rate of change accelerates, the time available to respond diminishes.

With accelerating change, shrinking product life cycles, increasing competition, overwhelming amounts of information, and persistent demand for higher quality and productivity, there is significant pressure on organizations to find new ways to manage.

With all of this new complexity and rapid change comes uncertainty, and with greater uncertainty there is simply less that managers can control. This leads to a critically important question. How do you manage when the capacity for control is reduced because of the environment?

Dr. Edwards Deming thought that management was prediction, by which he meant that every decision is also a prediction about the future. But how can you predict an increasingly (due to the rate of change in change itself) unpredictable future? You move from planning to preparation to enable flexibility and agility. You move from silos of expertise to collaborative co-creation of ideas, plans, products and services—with workers, partners, suppliers, customers and even competitors. You move from a hierarchical model of your business to one that looks more like a cloud.

In the evolving world of business, the structure, content, and process of work is changing:

- Less rote and process, more cognitively complex and creative
- Less structured and directed; more team-based and collaborative
- With knowledge readily available via technology, success is more dependent on social skills and the ability to share information constructively
- More dependent on technological competence in order to find, marshal and use the right resources at the right time

- More time pressured
- More mobile and less dependent on geography

Organizations will continue to be:

- Leaner and more agile
- More focused on identifying value from the customer perspective
- Less focused on planning and more focused on preparation in order to be agile
- Less hierarchical in structure, work and decision authority
- Continually reorganizing, breaking up, acquiring, disposing, partnering and increasingly looking at adjacent opportunities to maintain or gain competitive advantage

So what is driving all of this rethinking of the enterprise, whether intentionally or not, that is tightly coupled with, and being driven by, the Cloud and the need for innovation?

Let's zoom in from the view at 100,000 feet above the Earth and look at some of the current macro trends already affecting the workplace at ground zero on earth.

Macro Trends Affecting the Workplace

tinyurl.com/37o35rg

The New Normal
Cloud Computing Work Mobility
The Hollywood Model
Unexpected Economic Change
Regulations and Cyber Jurisprudence
A New Workforce / An Old Workforce
Risk and Security Management
Pervasive Simplicity
Innovating to Zero
IT Consumerization

Work Mobility and the Hollywood Model

Did you know that according to the U.N. telecommunications agency the number of mobile phone subscriptions worldwide had reached five billion in 2011? Did you know that these devices are how their users get to the World Wide Web? Did you know that their mobile phones will soon be able to surf the Web even if they aren't smart phones? IBM India Research Laboratory has developed the "Spoken Web," that parallels today's browser-based Web. People will talk to the Web and the Web will respond. For this technology, researchers have developed a new protocol—Hyperspeech Transfer Protocol (HSTP)—which is similar to the Hypertext Transfer Protocol (HTTP). With the earth this hyper-connected, the quaint notion of commuting to a workplace to work is so 20th century.

The incoming workforce is composed of digital natives and bring with them a whole new wave of change that requires reassessment of traditional supervision, control and work structures. They require information anywhere, anytime and on any device. They also require 7x24x365 access as described in the new work force below.

Work is not only mobile in the sense that it follows our workers around, work is also rapidly moving beyond corporate boundaries. More companies are pursuing the "Hollywood" business model (as described by Tom Peters in his book, *In Search of Excellence*). Back in the day, movie companies in Hollywood owned the studios, the actors and even the channels of distribution.

Not so today. Each new blockbuster movie is the equivalent of an IPO. Everything is a *project*. At the end of each project the players are disbanded. The key is to focus on where the company can best create value and outsource everything else to others, because those others are focused on where they can best create value and therefore will likely do the job better, faster and cheaper than you could. This provides dynamic scale and responsiveness while quarantining costs and keeping them variable.

A New Workforce / An Old Workforce

In addition to the technologies that are connecting the entire population of the planet, work mobility is also a reflection of socio-economic changes in the workforce. Profound demographic changes are occurring in the workplace.

Generation:	Traditionalists	Baby Boomers	Generation X	Millennials	Gen 2020
Born:	Before 1946	1946-1964	1965-1976	1977-1997	After 1997
Population:	46 M	78 M	50 M	88 M	41 M* * as of 2008
Defining Events:	WW II, Korean, Cold Wars, Suburbs	Watergate, Woodstock, Women's Rights, JFK	MTV, AIDS, Gulf War, 1987 Crash, Berlin Wall	Google, Social NWs, 9/11/2001, Obama	Social Games, Iraq War, Recession...
Defining Technology:	Mainframe Fax Machine	Personal Computer	Mobile Phone	Google & Facebook	iPhone & Android Apps

Adapted from "The 2020 Workplace" Meister & Willyerd

The Millennials (1977-1997) are so-called "Digital Natives." They are technologically savvy, hyper-connected and mobile. They communicate by using blogs, social networks, wikis and video sharing. They bring a different perspective to work. Their family, friends, and coworkers are constantly in touch, and straddle the fence of "*Weisure Time*," the blurred line between work and leisure.

According to this new work force, work is that brief period during the day where they have to use old, out of step, productivity interfering technology. The new workforce expects constant availability because they intermix work, home, play (and sleep) intermittently during the entire 24 hour day. Google, Amazon and Facebook, for the most part, has instilled in them an expectation of *good enough* to get the job done and impatience for waiting for *perfect*, and they have no tolerance for digital failure or systems that cannot work the way they do.

Source: wellingtongrey.net

But what about the old work force? They are retiring, they are forgetting, they are dying—and with them goes the knowledge and skills needed to preserve existing business models, processes and systems; not to mention the deep and abiding relationships they have among customers, suppliers, partners and themselves. This is causing what has been referred to as a "huge knowledge gap." As the current generation approaches its retirement years, companies need to start changing the ways that they define work. Assuming current downturned economic conditions are not permanent, boomers make up about one-third of the U.S. workforce, and there aren't enough younger workers to replace them. Labor shortages in key industries will force a radical rethinking of recruitment, retention, flexible work schedules and retirement.

Researchers Lynne Morton, Lorrie Foster and Jeri Sedlar say that "Today's linear life plan of distinct years for education, work and leisure is becoming obsolete. In its place is emerging a cyclical and phased life plan in which they exist in different proportions throughout life."

Companies need to develop ways for older workers to pass their knowledge and skills on to younger workers. They need to make a concerted effort to integrate older and younger workers. Consumer-oriented companies need to recognize that their customers are also aging and prospective clients often feel more comfortable interacting with an older representative and, therefore, they should make an effort to retain and hire those with extensive experience.

There are cultural impacts on losing the older work force, both positive and negative. Whether culture shifts are desirable or not, a large influx of new employees has the potential of changing the cultural underpinnings of an organization. An aging workforce preserves a company's culture and way of doing business that can be stable, but can also block new thinking. So the issue is not just knowledge retention into the new work force, but cultural preservation of those elements that drive success and characterize the company. This issue of balance is an imperative as companies increasingly outsource work that does not create customer value or competitive.

And there is the "wisdom" issue. Older workers are perceived to be better at interpreting facts and trends; a product of their high-value experience and expertise. Younger workers seem to accumulate facts and repeat them; a product of the "Google generation." Insights are important in an environment that is rapidly changing, especially considering that the pace of change itself is increasing. On the other hand, that same rate of change suggests risks of experience and expertise blinding the organization to what is really going on. The speed of change and the technological underpinnings of evolving business practice suggest a young, nimble workforce to keep up and stay abreast of what is happening everywhere. The key is to find a balance.

Organizations that do a good job of succession planning, talent development, emergent processes that workers create to make things work, not the formal processes some manager drew on a chalk board, and creating innovative ways to continuously and flexibly engage their older workforce will be ahead of the *aging curve.* Such firms will be on the way to higher retention rates, higher productivity and higher employee satisfaction, creating significant future financial and non-financial returns. Specific solutions will be as unique as the businesses themselves, accommodating employee demographics, business issues, competitive pressures, cost drivers, and financial and business risks.

The enabling technologies behind work force innovation include the Cloud, social networks in the Cloud and ubiquitous bandwidth. The impact on traditional organizations is huge, affecting corporate culture and hiring practices. Further, there's no going back and no place to run and hide.

With proper governance, smart companies will expand the use of social networks. They'll deepen their understanding of the implications of online personal behavior and its interaction with employees, customers and suppliers. Central to going forward is the need to address business process management and organizational structures to meet younger employee expectations and balance individual choice with information security.

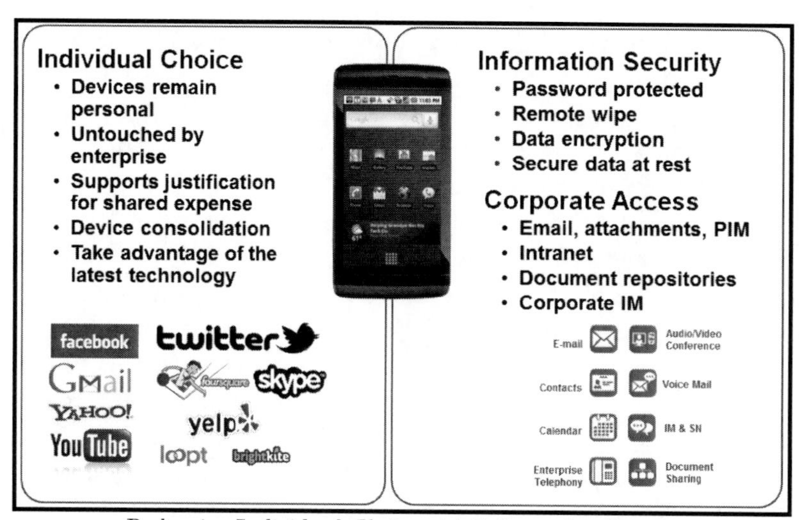

Balancing Individual Choice with Information Security

The New Normal: IT Consumerization

If we look at the macro-trend of IT Consumerization, we'll see that the introduction of *new* technologies has shifted from the workplace directly to consumers. This has been facilitated by the ease of use of Web 2.0 technologies, economies of scale and lower hardware costs, pushing early adoption not only by so-called Digital Natives, but also the rest of us. Today, even Grandma keeps up with her scattered family by using Skype, Facebook, Yahoo mail and Twitter. As Apple has taught us, ease-of-use, simplicity and taste are the determining factors of adoption.

Another facet of IT Consumerization is the increase in employee-owned assets at work, smart phones, netbooks, and tablets. This Bring Your Own Device (BYOD) trend offers organizations many advantages including increased productivity, lower capital and support costs as these costs are shifted to employees. On the other hand, there are many barriers for business and government organizations:

- Psychological
- Difficulty justifying "Toy" versus "Tool"
- Security
- Digital Rights Management
- Ownership and support issues not clearly defined
- Entitlement and the challenge to distinguish between personal uses and business uses of the equipment
- Operational changes need to support diverse employee-owned devices and software

We are now halfway to somewhere known as the digital revolution. That is to say, digital is no longer *special*, it's now just *normal!*

Although we have already gone through a lot of change, what lies ahead of us involves you and me. The New Normal will have an enormous impact on the way companies organize and communicate with customers—and the way they have to be organized internally.

Organizations are increasingly faced with customers and consumers who no longer tolerate limitations in terms of pricing, timing, patience, depth, privacy, convenience and intelligence. A number of new rules apply to The New Normal:

- Consumers will have zero tolerance for digital failure.
- Good enough beats perfect
- The era of total accountability
- Abandon absolute control
- Every interaction with a customer must be viewed as a make-or-break moment for the relationship with the customer.
- IT departments will have to adopt a new way of working. They no longer have to build pyramids; they must put up tents instead.
- Companies have to adapt their information strategy, their technology strategy, their innovation strategy and the way they are organized internally.

Another aspect of the consumerization of IT is the loss of control of *architecture*. In the beginning , the government basically funded the R&D that drove IT, and thereby dictated the nature of the equipment that IT had to work with to deliver its value. Then as business began to be the primary revenue stream for technology companies, the enterprise began dictating what technology looked like and how it was used. We have now entered the stage where consumers and hyper-scale Cloud providers, not enterprises, are dictating the R&D to meet their needs and desires for future computer architectures. Corporate IT will have to adapt instead of dictate as in the past.

The New Normal: Unexpected Economic Change and Innovating to Zero

Regardless of the organization—public, private, business, government, or even individuals—you can detect an emerging super-ordinate goal to innovate to zero. It may or may not be explicitly expressed, but there is some metric that is desired to be moved closer and closer to zero over time. It might be zero carbon, zero waste, zero consumption, zero time, zero costs, zero staffing, zero defects—there is some metric the person or organization is trying to move to zero.

Watch Bill Gates: Innovate to Zero

tinyurl.com/7fz36n4

One effect of this drive to zero is that the global economy is being increasingly driven by extreme volatility characterized by slow growth; high unemployment; people borrowing less; saving more while spending with greater caution; and high levels of economic insecurity as everyone tries to figure out (intuitively or explicitly) what an economic model based on zero costs, zero waste, zero input, zero time to deliver, and so on really means. As a result, businesses are becoming more adverse to making capital expenditures and extremely cautious in increasing operating expenses. They are strapped with tight budgets and are grappling with correlating costs to revenues.

It's no wonder businesses are placing a renewed focus on how they can gain advantage in the new world of unexpected change.

Writing in the *Harvard Business Review* in 2008, John Hagel, John Seely Brown and Lang Davison commented that, "The historical pattern disruption followed by stabilization has itself been disrupted. A new kind of infrastructure is evolving, built on the sustained exponential pace of performance improvements in computing, storage, and bandwidth. Because the underlying technologies are developing continuously and rapidly, there is no prospect for stabilization … making equilibrium a distant memory."

Punctuated equilibrium has itself been punctuated.

Cloud Computing: A Technology-Enabled Game Changer

Cloud computing is not a technological breakthrough as much as it is a new way to deliver economic value. It merely provides an innovative computing platform to transform the way IT delivers value to the businesses and individuals. Cloud computing represents a major transformative shift from on-premise IT services to on-demand, "self-service" IT delivered over the Internet, fostered by a kaleidoscope of converging events, trends, and shifts. Besides technology, there are other factors at work spanning the social, economic, environmental, and generational shifts that are now underway.

As the Cloud era is only beginning to emerge, we can't currently envision the new processes and business models it will enable. However, it is worth looking at what is happening now. Consumers and businesses are using Cloud services for:

- Accessing and downloading media.
- Accessing and downloading mobile apps.
- Accessing and running business applications (CRM, e-commerce, logistics, provisioning, etc.).
- Collaborating with colleagues, clients, and customers (project management, online communities, email, meetings).
- Analyzing immense amounts of data (Big Data).
- Storing large amounts of data (much of it unstructured, like video, images, text files, etc.).
- Developing and testing new applications and online services.
- Running distributed applications that need high performance. (All of the social media apps we use are essentially Cloud applications—they run on virtual machines hosted in mostly 3rd-party data centers all over the world).
- Scaling IT operations to handle seasonal and other peak-load requirements to take advantage of buying computing capabilities by the hour, rather than pre-paying for capacity rarely needed.
- Back up and Disaster Recovery—keeping copies of our systems and data in remote locations, ready to run if a natural disaster impacts our normal operations.

In short, cloud computing in all of its instantiations—Software as a Service, Platform as a Service, Infrastructure as a Service and "Everything as a Service"—is here to stay. Taking advantage of the Cloud is the most scalable and the most cost-effective way to provide computing resources and services to anyone who has reliable access to high bandwidth networking via the Internet.

We're now committed to living in the mobile Internet era. We treasure our mobility and our unfettered access to information, apps, media, and services. Small businesses and innovative service providers have embraced cloud computing and services wholeheartedly and are already reaping the benefits of *pay-as-you-go* software and computing and storage services.

Medium-sized businesses are the next to embrace cloud computing, because they typically don't have the inertia and overhead that comes with a huge centralized IT organization.

Large enterprises' IT organizations are often the last to accept cloud computing as a safe and compliant alternative for corporate IT. Yet many departments in those same large enterprise organizations have been the early adopters of cloud computing for the development and testing of new software products and for the departmental (or even corporate) adoption of SaaS for many of their companies' most critical applications. This is called "shadow IT."

In light of the New Economic Normal, businesses want IT as a utility, metered, available on demand, with unlimited supply. They want to shift from capital expenditures (CAPEX) to operational expenditures (OPEX) that are tied *elastically* to the ups and downs of business activities. They want to buy *computing*, not *computers*.

Utility computing and on-demand computing have been around for quite some time, and serve as foundations of cloud computing. Geva Perry, chief marketing officer at GigaSpace Technologies, explains, "The main benefit of cloud computing is better economics. Corporate data centers are notoriously underutilized, with resources such as servers often idle 85 percent of the time. This is due to over provisioning—buying more hardware than is needed on average in order to handle peaks (such as the opening of the Wall Street trading day or the holiday shopping season), to handle expected future loads and to prepare for unanticipated surges in demand. Utility computing allows companies to only pay for the computing resources they need, when they need them."[1]

A report in *The Economist* expands on issues related to corporate data centers, "Before Ford revolutionized car making, automobiles were put together by teams of highly skilled craftsmen in custom-built workshops. Similarly, most corporate data centers today house armies of 'systems administrators,' the craftsmen of the information age. There are an estimated 7,000 such data centers in America alone, most of them one-off designs that have grown over the years, reflecting the history of both technology and the particular use to which it is being put. It is no surprise that they are egregiously inefficient.

"On average only 6% of server capacity is used, according to a study by McKinsey, a consultancy, and the Uptime Institute, a think-tank. Microsoft's data center in Northlake, just like Henry Ford's first large factory in Highland Park, Michigan, may one day be seen as a symbol of a new industrial era."[2]

In the Cloud, everything is delivered as a service (XaaS): Infrastructure as a Service (IaaS), Platform as a Service (PaaS) and Software as a Service (SaaS) as shown in the figure below.

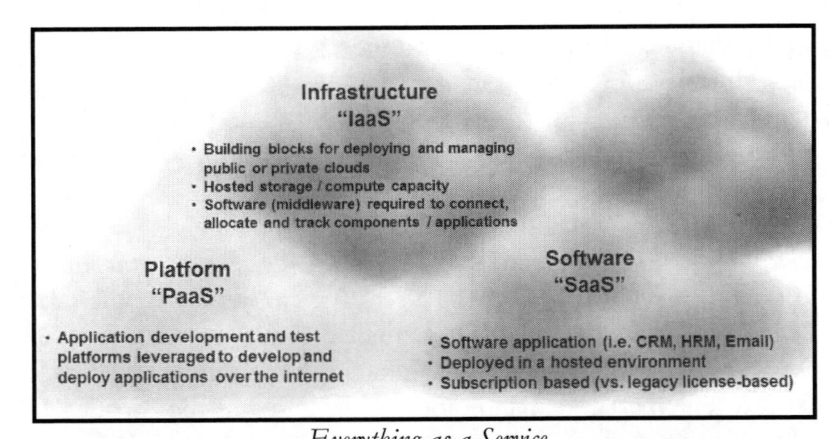

Everything as a Service

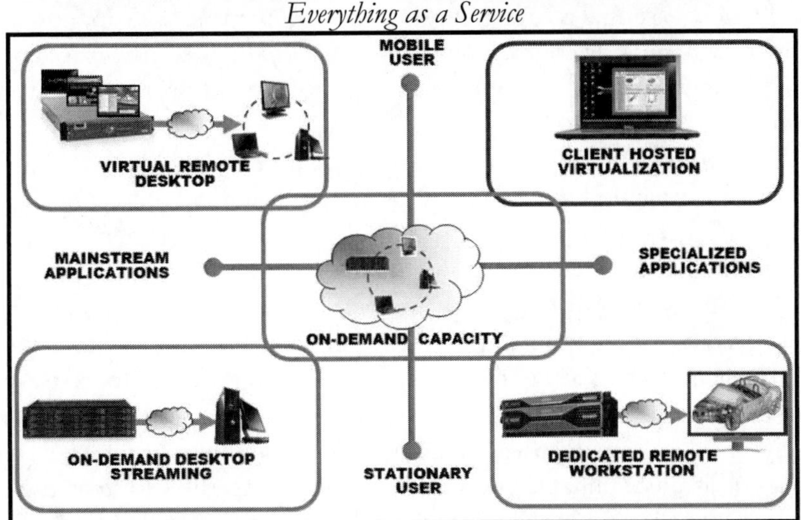

The Cloud is far more than the use of remoter servers and services. It's all about flexible new computing models for the users of information technology in all its forms, from virtual desktops to on-demand capacity, regardless of device, location or time of need.

Again, what is the Cloud? In one sense, there's no such thing as cloud computing. As we stated earlier, it's not a new architecture; it's not a new technology; it's not a new methodology. It is however, a radically new delivery model. Cloud computing enables dynamic provisioning of computing resources over the Internet.

In short, the Cloud *is* the computer.

The central business driver is "rapid elasticity," meaning that, on-demand, users can scale up or down, with no up front costs, havig their usage metered in a pay-as-you-go costing model.

Cloud Computing: A Business Model Game Changer

Projects, ideas, dreams and world-changing thoughts can all be launched using cloud computing. The Cloud truly makes for a flat, resource un-constrained world. The only limit is your imagination, so think of cloud computing not as a stand-alone skill or feature but as an integral part of the workplace that workers of the future are going to encounter.

As cloud computing goes mainstream, the barrier to conceiving new ideas and bringing them to the customer will go down dramatically. This is the sweet spot that innovative application developers should be targeting. CIOs are eager to see if the potential of the Cloud can be transformed into tangible business value that will help them deal with three extremely pressing issues:

- *Moving Faster:* Every organization these days is looking to accelerate its pace of business to be able to respond to shifting market demands and opportunities as they occur, not 6-9 months later. And with IT now essential to every facet of those organizations' operations, it's no longer acceptable to be told that major company initiatives will be bogged down for months because that's how long IT needs to spec out, purchase, test, set up, and provision new systems. Cloud deployments breed speed.
- *Lower The Cost Of Infrastructure:* Although companies are demanding more and better IT services and performance, the available budgets for CIOs are barely above the viciously scaled-back levels of 2009. So CIOs are hoping a thoughtful Cloud strategy will let them bridge the yawning gap between expectations and demands, and limited funding.
- *Flipping the 80/20 Ratio:* Many CIOs say they are still spending 70 or 80 percent of their IT budgets on maintenance versus innovation, or on run versus build. As a result, very little money is left in the IT budget to fund new and vital customer-facing projects because it all gets sucked up in keeping the lights on. In turn, this funding trap is causing many CEOs to want to figuratively shoot their CIOs, or outsource all IT operations. Regardless of which of those choices is made, the status quo is unsustainable and CIOs are looking to the Cloud as the proverbial magic bullet that they can aim at reversing the 80/20 status. But the real value in the Cloud is the level of innovation it will enable.

A really big game changer of cloud computing for businesses is that they will need to make only minimal hardware and software investments to achieve a new level of IT cost savings as the entire spectrum of business technologies and services becomes accessible in the Cloud. But it won't be just cost savings. After all, a company cannot save its way to market dominance. In this sense, "IT Doesn't Matter," but "Business Innovation Processes Do."

The Cloud makes it possible to create new "business operations platforms" that allow companies to change their business models and collaborate in powerful new ways with their customers, suppliers and trading partners—stuff that simply wasn't practical before.

Watch : *tinyurl.com/7nt5mos*

> The real value of the Cloud
> isn't about on-demand technology,
> it's about on-demand business innovation.

New, distributed computing architectures are changing the ways companies develop "computer applications." Using Web services, service-oriented architectures (SOA), and choreography languages such as the Business Process Execution Language (BPEL), systems developers can rapidly *mashup* services to create new functionality.

Cloud computing is a different model built around *services* and not *applications* or *infrastructure*. And those services are "units of business" (e.g., calculate discount, lookup customer credit rating, and so on), not "units of technology" (e.g., servers, storage, bandwidth, and so on). For example, a new Order Entry system can be developed by mashing up "Cloud services" from existing enterprise systems provided in-house and by third party suppliers.

Over time, as more and more Cloud services are developed, companies can avoid the total re-writing of existing applications to fit the Cloud. But even more critical for companies is the pressing need to choreograph and then manage complete end-to-end business processes in multi-company Business Networks. They want to manage "situational business processes" for rapid innovation that spans multi-company value chains. Those situations include new programs, campaigns and initiatives, or responding to competitive threats.

The book, *Enterprise Cloud Computing: A Strategy Guide for Business and Technology Leaders,* goes beyond composite applications and on to Situational Business Processes, "While a composite application might be designed to do a specific task or handle a specific activity for a unique situation, the *choreography* and management of end-to-end business processes that make up any given value delivery system goes one giant step further. And that's where Business Process Management (BPM) comes in. In other words, composite applications are a great advancement in delivering software applications, while Situational Business Processes are a great step forward in delivering business processes that tap and coordinate the process fragments contained in applications. And the governance inherent in the use of the BPM system brings management control to 'software gone wild' possibilities with Situational Apps."

"The bottom line is that business process management is the foundation for going beyond 'mashup applications.' Companies don't want more software applications, they want business processes that deliver value to their customers, and such business processes are no longer 'owned' by a single company, no matter how dominant that company is in the value chain."

Business Process Management (BPM) is what sets "enterprise cloud computing" apart from "consumer cloud computing." Because the average end-to-end business process involves over 20 companies in any given value chain, multi-company BPM is essential to business innovation and maintaining competitive advantage. Bringing BPM capabilities to the Cloud enables multiple companies to share a common BPM environment and fully participate in an overall end-to-end business process.

BPMaaS can be implemented as a "horizontal" Business Operations Platform (BOP) that has a Business Process Management System (BPMS) at its heart. Call this the new "business operating system" of the new Business Network. We'll discuss business process management in the Cloud further in Part 2 of the book.

Risk and Security Management: Regulations and Cyber Jurisprudence

The key issues for adopting any advanced technology are fear, uncertainty and doubt. Why? What's kind of cool to observe is that the *first principles* change little, even as they get embedded in ever more sophisticated technology, whether it's going from paper journals to cash registers, or from cash registers to point-of-sale systems. Each step change in technology represents a two-edged sword—more power and more loss exposures. Any organization faces loss exposures, and the first principles of security, auditability and management controls must be adapted to new technologies.

What's the key issue related to adopting cloud computing? You guessed it, security.

Today's concern over security has several dimensions. Currently, technology addresses security through point solutions, point processes and physical security. Standards are "nice to have" and interoperability between hardware and software opens security gaps. Governments constantly have to balance economic risks and benefits and cope with jurisdictional conflicts and geopolitical boundaries in cyberspace. Companies have to deal with legal issues and competitive threats. Meanwhile, individuals need to be rational and prudent as they deal with the likes of Facebook and online financial services.

"Any sufficiently advanced technology is indistinguishable from magic."
—Sir Arthur C. Clarke, Author of *2001: A Space Odyssey,*

So what does the future portend for security? A new security architecture is needed that includes hardware, software and development environments based on Digital Rights Management (DRM)) in order to support the many rules and regulations that are coming into play for compliance on data location, access and usage. Further, as shown in the technical table below, no one "identity management and authentication program" is appropriate for all Internet uses.

Hardware	Software
▪ Policy Information Points ▪ Location services (GPS) ▪ Policy Enforcement Points ▪ Biometrics ▪ Bluetooth phones ▪ Encryption points services ▪ Secure flash ▪ CPU keys	▪ Security Assertion Markup Language (SAML) ▪ eXtensible Access Control Markup Language (XACML) ▪ Hashed binaries ▪ Pedigreed binaries ▪ Stateless sessions

Much work is currently underway by industry groups such as the Cloud Security Alliance (cloudsecurityalliance.org) and the Object Management Group's Cloud Standards Customer Council (cloud-council.org) to place a stake in the ground for Cloud security standards. As that work advances, it's useful to note that unless your organization has the resources to staff world-class security experts, security often becomes a major reason for choosing third-party Cloud service providers for they do indeed employ the very best security technologists. Because almost every company on the planet has some sort of Internet presence, security may in fact be a major advantage instead of a major fear factor of cloud computing.

The rise of many new jurisdictional regulations on data location, access and consumption are generating a compliance nightmare for many organizations. Think about it, we have seen this movie before—it was called PAYROLL. Very few companies do their own payroll anymore—not because it is difficult to do, but because most organizations cannot afford all of the lawyers and accountants needed to keep up with all the rules and regulations in order to stay in compliance, not to mention all of the people to keep the systems up to date once the rules and regulations are understood. Like payroll rules, this is likely to force many companies to turn over (move to the Cloud) their management of data and information.

Smart Everything + Pervasive Simplicity = Massive Computing Needs = The Cloud

> Think about it; a teenager with little to no technical knowledge can press one icon on her smart phone and consume more computing resources than were used in putting a man on the moon.

That is the power, and cost, of the path that *pervasive simplification technology* has put us on. Not only is it easy for a person to now request either directly or indirectly massive amounts of computations, but the learning barriers to do that have dropped so low that the number of people who can do it almost extends to the entire world population.

Think about it. Ask SIRI a question today on your iPhone and you set in motion processes that consume as much if not more compute, network and storage resources that would have required an entire team of expensive programmers and millions in dedicated hardware, not to mention days to get the response. Instead, it happens instantaneously. Plus, it was so easy, you don't remember the result. So later you just ask again.

Add to that the fact that every device in our lives is becoming smart—our phones, our cars, our refrigerators, our toilets—and in the process generating so much data that the ability of a human to find and utilize it becomes constrained.

This means several things. First we need massive amounts of storage that are beyond the capacity of single organizations. Only by sharing in the Cloud can we benefit from all this data. Also, it will take massive amounts of CPU cycles to analyze this data, so much so that the ability of an organization to have that excess (used once in a while) capacity is compromised. This means most, if not all, analytics will move to the Cloud. This also means new computer architectures that break the sequential bonds of the classical Von Neumann architectures. Say hello to the cognitive computer—coming to a Cloud near you.

The Cognitive Computer

Imagine that immediately as new technologies and capabilities come online in advanced laboratories, they could be made available to world at large. That is one of the potentialities of the Cloud. And once those capabilities are available, then the forces of innovation can be brought to bear. Case in point—the cognitive computer.

There are 100 billion neurons with 1,000 trillion connections in the human brain. Scientists at Manchester University (England) are working toward producing a high-performance computer that they intend to use to create working models of human brain functions. Microprocessor chips will be linked together to simulate the highly-complex workings of the brain, whose functionality derives from networks of billions of interacting, highly-connected neurons. The "chips" used for this project are based on ARM processor technology (a chip based on the ARM technology is almost certainly inside your smart phone as you read this) and will form the system architecture for a massive computer, called SpiNNaker (Spiking Neural Network architecture). The initiative, led by Dr. Steve Furber of the University of Manchester, hopes to map out the brain's individual functions. Spinnaker is essentially an *artificial neural network* realized in hardware, a massively parallel processing system eventually designed to incorporate a million ARM processors.

Meanwhile, by replicating the functions of neurons, synapses, dendrites and axons in the brain using special-purpose silicon circuitry, IBM claims to have developed the first custom cognitive computing cores that bring together digital spiking neurons with ultra-dense, on-chip, crossbar synapses and event-driven communication.

IBM's effort is sponsored by the Defense Advanced Research Project Agency (DARPA) to build Systems of Neuromorphic Adaptive Plastic Scalable Electronics (SyNAPSE). IBM and its university partners, Columbia University, Cornell University, and University of California-Merced, and the University of Wisconsin-Madison, have the eventual goal to create a brain-like cognitive computer with comparable size and power consumption to the human brain.

"We want to extend and complement the traditional von Neumann computer for real-time uncertain environments," said Dharmendra Modha, the principal investigator of the DARPA project and project leader for IBM Research. "Cognitive computers must integrate the inputs from multiple sensors in a context dependent fashion in order to close the real-time sensory-motor feedback loop."

Traditional von Nuemann computers are ill-equipped to deal with the multiple simultaneous data streams coming in from sensors today (e.g., Internet of Things), but brains handle these easily by distributing processing and memory among its neural networks. Modha said, "This is the seed for a new generation of computers, using a combination of supercomputing, neuroscience, and nanotechnology. The computers we have today are more like calculators. We want to make something like the brain. It is a sharp departure from the past."

Dr. Steve Furber, *Dr. Dharmendra Modha*

Watch Dr. Modha's Keynote
tinyurl.com/77p9659

If it eventually leads to commercial brain-like chips, the project could turn computing on its head, overturning the conventional style of computing that has ruled since the dawn of the information age and replacing it with something that is much more like a thinking artificial brain. The eventual applications could have a huge impact on business, science and government. The idea is to create computers that are better at handling real-world sensory problems than today's computers can. IBM could also build a better Watson, the computer that became the world champion at the game show Jeopardy!

This new computing unit, or core, is analogous to the brain. It has *neurons*, or digital processors that compute information. It has *synapses* which are the foundation of learning and memory. And it has axons, or data pathways that connect the tissue of the computer.

These new chips won't be programmed in the traditional way. Cognitive computers are expected to learn through experiences, find correlations, create hypotheses, remember, and learn from the outcomes. They mimic the brain's "structural and synaptic plasticity." The processing is distributed and parallel, not centralized and serial.

With no set programming, the computing cores that the researchers have built can mimic the event-driven brain, which wakes up to perform a task. The IBM team has tested the chip on solving problems related to navigation, machine vision, pattern recognition, associative memory (where you remember one thing that goes with another thing) and classification. This is precisely the kind of technology that will be needed to power Big Data and predictive analytics that we discuss later in the book.

Eventually, IBM wants to combine the cores into a fully integrated system of hardware and software. According to Modha, IBM wants to build a computer with 10 billion neurons and 100 trillion synapses. That's as powerful as the human brain. The complete system will consume one kilowatt of power and will occupy less than two liters of volume (the size of our brains), Modha predicts. By comparison, today's fastest IBM supercomputer, Blue Gene, has 147,456 processors, more than 144 terabytes of memory, occupies a huge, air-conditioned cabinet, and consumes more than 2 megawatts of power.

As a hypothetical application, IBM said that a cognitive computer could monitor the world's water supply via a network of sensors and tiny motors that constantly record and report data such as temperature, pressure, wave height, acoustics, and ocean tide. It could then issue tsunami warnings in case of an earthquake. Or, a grocer stocking shelves could use an instrumented glove that monitors sights, smells, texture and temperature to flag contaminated produce. Or a computer could absorb data and flag unsafe intersections that are prone to traffic accidents. Those tasks are too hard for traditional computers.

"If this works, this is not just a 5 percent leap," Modha said. "This is a leap of orders of magnitude forward. We have already overcome huge conceptual roadblocks."

Indeed, this advancement of the Cognitive Computer on a Chip will recreate the very foundation of how computation happens—and what the very term, *computing,* actually means.

Likewise, the opportunity to transform these capabilities into innovations (products, services, business processes and business models) is probably beyond immediate imagination. It also highlights again the importance of the Cloud. While capabilities like this will be expensive to individually acquire, they can become part of the Cloud and available to all at equal cost that can be mapped directly to generated value, once again leveling the playing field for creating innovations.

Windows to the One

So, is it time to close the patent office? After all we now have smart phones and intelligent tablets. What else do we need to invent?

For a start, it's likely that just about everything may become an intelligent user interface with advances in "surface computing:" multi-touch computers, tabletop computers, window computers and the "work wall." All these devices, including the human body will be a window into "the One" (coined by Kevin Kelly), the intelligent World Wide Web.

Now, just as today's tablets were essentially designed at the legendary PARC labs under the direction of Alan Kay way back in 1968, the ideas of "everything as a computer" and "everything as a human interface" aren't new. Just take a glance at "Connections: AT&T's Vision of the Future" produced way back in 1997.

vimeo.com/10541443 or tinyurl.com/6m85kvg

The human body as an interface? Just check out the joint research of Carnegie Mellon University's Human-Computer Interaction Institute and Microsoft Research proof-of-concept implementation of OmniTouch. Results suggest interactions traditionally reserved for dedicated touch surfaces can now move into the environment in an ad hoc fashion. Although their current prototype is fairly large, there are no significant barriers to miniaturization. It is entirely possible that a future incarnation of OmniTouch could be the size of a box of matches, worn as pendent or watch. Thus, the benefit of extreme portability can be combined with the ease and accuracy of interaction on large, physical surfaces.

The chief goal of the project is to demonstrate that touch input can be achieved on everyday surfaces, including the human body. This brings to reality many intriguing interactions proposed in earlier work. There are also many opportunities to enhance the approaches in the current work. For example, it is possible to create 3D meshes of objects, allowing for distortion free projection onto non-planar surfaces (i.e., projective texturing). Additionally, postures and gestures of the arms and hands could be used for input. For example, a "telephone" gesture with hands could summon a keypad on the arm, while a "back of hand" pose could render a watch on the wrist.

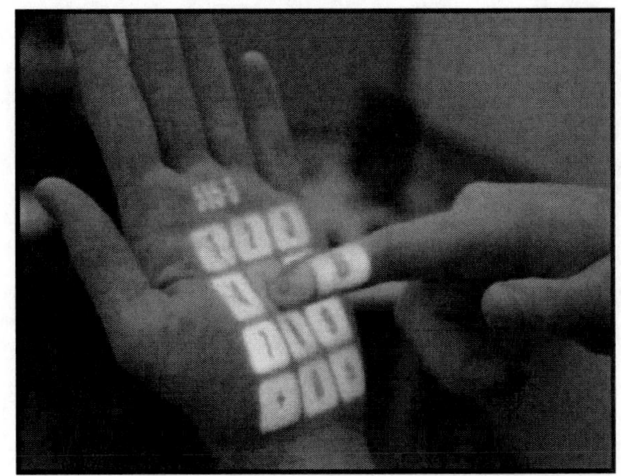

chrisharrison.net/projects/omnitouch/omnitouch.pdf
or tinyurl.com/89fk2g6

Finally, there are many fascinating qualitative questions surrounding on-body interaction. For example, how comfortable are people using their own bodies as interactive platforms? What are the social implications if other people want to use these interfaces? Is it acceptable to control an interface on someone else's body? Such interactive capabilities may open new questions in proxemics research.

Today, the amount of money being spent on research for military robotics surpasses the budget of the National Science Foundation, which, at $6.9 billion a year, funds nearly one-quarter of all federally supported scientific research at the nation's universities. The development of a new generation of military robots, including armed drones, may eventually mark one of the biggest revolutions in warfare in generations.

Throughout history, from the crossbow to the cannon to the aircraft carrier, one weapon has supplanted another as nations have strived to create increasingly lethal means of allowing armies to project power from afar. A model of an insect-size US Air Force drone is held by a member of the Micro Air Vehicles team of the Air Force Research Laboratory at Wright Patterson Air Force Base in Dayton, Ohio. Their goal is to make drones so small that they resemble small birds and insects – some will have wings that flap – and so sophisticated that they can operate in complex urban environments.

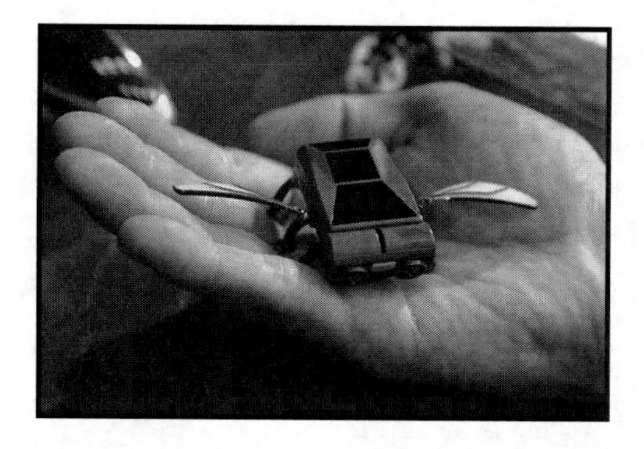

tinyurl.com/6qgn4lc

While there is speculation in some of these technologies discussed, the point is that they, and others, offer tremendous opportunity for innovations. The nature of the Cloud is that once they are available, they can be made available to all organizations as capabilities they can call on to enable their innovation initiatives.

Takeaway: Putting the Macro Trends All Together

Taken together, the macro trends we've discussed represent an exponential change curve that has just taken off like a rocket. As we have no prior framework, we must begin anew to build up the tacit information we need to apply to each component, and to the whole.

Each decade or so it seems there is a seismic shift in information technology. Such shifts change not only the impact on business capabilities, but also which technology companies become the leaders. Who would have thought that a retailer would become one of the first mega players of cloud computing? Once thought of as a book store, Amazon is a leading technology contender in the Cloud.

> What people do creates technology;
> then, technology enables what people do next.

References

---

[1] http://gigaom.com/2008/02/28/how-cloud-utility-computing-are-different/

[2] http://www.economist.com/specialreports/displayStory.cfm?story_id= 12411882

4. What Exactly is Business Innovation?

"Research is the transformation of money into knowledge. Innovation is the transformation of knowledge into money."
—Dr. Geoffrey Nicholson, 3M, Inventor of the Post-It Note

"If I had asked the public what they wanted, they would have said a faster horse."
—Henry Ford

To fully understand this thing we call innovation, we need to fully understand this thing we call technology. Technology is the making, usage, and knowledge of tools, machines, techniques, crafts, systems or methods of organization in order to solve a problem or perform a specific function. By this definition it is technology that enables, facilitates and accelerates innovation.

To understand innovation we need to understand technology in the same way that to understand a dolphin we would need to understand the ocean.

Humankind and technology have a symbiotic relationship, each feeding off of and pushing the other in never-ending cycles. Humans create technology; technology creates what we do as humans. Then the cycle repeats. Each cycle can be called an innovation. Man shapes technology—then technology shapes man.

Okay, let's agree for a moment that the future is fully dependent on mankind's relationship with technology. Some (e.g., Ray Kurzweil in *The Singularity Is Near*) postulate that the fusion of mankind and technology will reach the point of "singularity."

Many readers of this book might consider themselves to be "technologists." But wait. What the heck is technology, technology of any sort, including Information Technology (IT)? Actually that's a huge question, and one that we can turn to two people who have studied this question in great depth, Brian Arthur, and Kevin Kelly.

and

Visit their Web sites ...

tuvalu.santafe.edu/~wbarthur

kk.org

Get back to us after you have digested their remarkable insights! But for our current discussion, let's just say that the relationship of technology (whatever that *is*) and society is huge. More powerful than religion or politics, it's technology that shapes societies and humankind in general. Consider the harnessing of fire, the invention of the alphabet, the wheel, the steam engine, mass production, mass customization, the airplane and the Internet.

Brian Arthur's *Nature of Technology* is an elegant and powerful theory of the origins and evolution of technology. It accomplishes for the progress of technology what Thomas Kuhn's, *The Structure of Scientific Revolutions* did for scientific progress. Arthur explains how transformative new technologies arise and how innovation really works. Conventional thinking ascribes the invention of technologies to "thinking outside the box," or vaguely to genius or creativity, but Arthur shows that such explanations are inadequate.

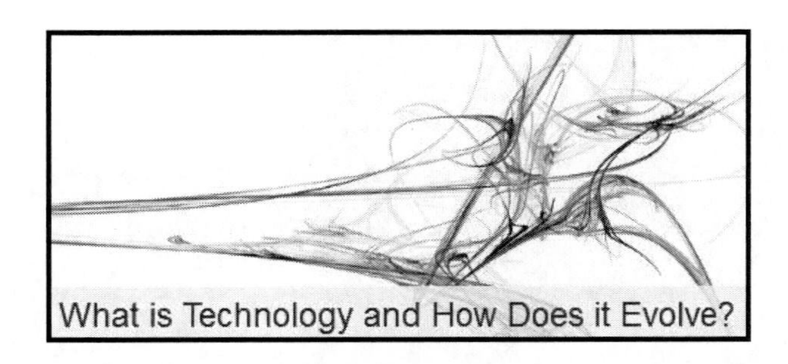
What is Technology and How Does it Evolve?

Rather, technologies are put together from pieces—themselves technologies—that already exist. Technologies therefore share common ancestries, and combine, morph, and combine again, to create further technologies.

Technology evolves much as a coral reef builds itself from activities of small organisms—it creates itself from itself (autopoiesis); and all technologies are descended from earlier technologies.

Arthur postulates that all future technologies will derive from those that now exist (perhaps in no obvious way) because these are the elements that will form further elements that will eventually make these future technologies possible.

"Creativity is just connecting things."—Steve Jobs

"Steve Jobs has created new ideas precisely because he has spent a lifetime exploring new and unrelated things—seeking out diverse experiences. Jobs hired people from outside the computing profession; he studied the art of calligraphy in college, meditated in an Indian ashram, and evaluated The Four Seasons hotel chain as he developed the customer service model for the Apple Stores. Look outside your industry for inspiration."[1]

Kevin Kelly takes it a step further, defining technology as *part* of our thought process, as a natural extension of evolution and a natural, living system. In fact, technology is not a sufficient word for what Kelly is trying to grasp, so he's coined a new word—the *technium* to mean a whole system (as in "technology accelerates"). He reserves the term technology to mean a specific technology, such as radar or plastic polymers. For example, he would say, "The technium accelerates the invention of technologies." In other words, technologies can be patented, while the *technium* includes the patent system itself. He then traces the technium back to our hunter-gatherer past to document how it evolved to its current state.

The technium gains immense power not only from its scale but from its self-amplifying nature. One breakthrough invention, such as the alphabet, the steam-pump, or electricity, can lead to further breakthrough inventions, such as books, coal mines, and telephones. These advances in turn lead to other breakthrough inventions, such as libraries, power generators, and the Internet.

Watch Kevin Kelly, The Next 5,000 Days of the Web
tinyurl.com/l77pgu

The technium is "human extended," something we've been building and evolving since we have been building and evolving. It is the "extended body for ideas." And to be in business today, to truly innovate, it is important to have a grasp on that "extended body for ideas"—to know that the answer to your business's troubles isn't to get on Twitter, but to understand the changing environment that bred it. You won't find any immediately applicable business solutions in *What Technology Wants*, but you will find something that could be much more important to your business—a new way of looking at the world. Here are some of Kelly's high points:

- Web's first 5,000 days
  - People expected "TV, only better"
  - Impossible to imagine Wikipedia, Facebook
  - Economic models
- December 2007
  - 100 billion clicks per day
  - 294 billion emails sent daily
  - 55 trillion links
  - 255 exabytes of magnetic storage
  - 5% of global electricity consumption
- Magnitude equivalent of a human brain
- Web's next 5,000 days
  - Doubling every two years
  - 6 billion human equivalents by 2040
  - Mobility
  - Digital universe fuses into physical world

– Our devices are windows into the Web ("The One")
– Internet creating a "global brain"

One could argue that Kelly did a pretty good job of forecasting, because in December 2010 there were, 2 billion users, 107 trillion emails sent, 47 billion text messages per day, 35 billion "client" devices (5 billion phones), 13 billion indexed pages (est. half of the total is static) and over 1 trillion dynamic pages. Please note these are approximations as precise hard numbers are hard to get and always subject to argument.

Kelly suggested there would be three outcomes associated with the growth of the Web. These included:

1. *Embodying the machine:* the Web would extend out into the real world more and more and would physically manifest itself as part of the world. We see this in many of the location aware apps today.
2. *Restructuring the architecture:* at the lowest levels how the machines talk to each other would need to change. Many of the newer de facto and de jure standards for Web and mobile enabled applications are examples of this as well.
3. *Codependence on new technology:* In the same way the introduction of the alphabet and writing. As with the expansion of literacy began the reduction of the memory ability and utilization of humans, we will become increasingly dependent upon this technology. How many people remember phone numbers anymore?

Kelly thinks that we will see the emergence of a Global Brain that will be smarter than the technology we are used to (IBM's Watson anyone?). The Web could become the ultimate man or woman "Friday"—that ever present assistant who always knows what you want and has it (information, reservations, locations, etc.) available to you, sometimes even before you ask for it. However, Kelly introduces one criterion to enable the Web to be all that it can be. Individuals must become *transparent* to gain those benefits. Many think that will not happen due to privacy concerns. But one only has to look at the growth of Facebook and location enabled "apps" to see that we have become very comfortable becoming transparent. Much work is underway to overcome the constraints of privacy versus transparency (maybe then you might go ahead and get a Facebook account).

Let's explore some modifications to some of his observations:

- *Disembodying information:* Big Data (the consequence of smart devices everywhere) will mean it will become increasingly uneconomical for organizations to hold all the information they need inside their fire walls. In fact, it will become increasingly hard, perhaps even impossible to point to a physical device and say, "That is where my data is." Data will become disembodied.
- *Restructuring us:* A quick look at augmented realty and the evolution of location-enabled apps on smart phones shows the beginning of our evolution toward becoming *cyborgs (tinyurl.com/7xzet)*.

- *Multisensory computing:* Using the entirety of the brain's ability to digest, integrate and project information requires these augmented multi-sensory models, with the visible endpoint looking something like the AlloSphere at the California NanoSystems Institute *(tinyurl.com/86e4bgy).* Visualizing, hearing and exploring complex multi-dimensional data provides insight that is essential for progress in a business of the 21st century, where the amount and complexity of the data overwhelm traditional computing environments. And if this seems too "out there," remember that the basic capabilities of the Allosphere can be rendered by a Nintendo 3DS.
- *The Pervasive Cloud:* Information will be Pervasive – there where we need it, when we need it. Think about the GPS enabled applications in smart phones today and their ability to provide information about your surroundings (who else is there, what is the best restaurant, how close is it, how do I get to it, what is on the menu, where is an ATM along the route, etc.) without you having any knowledge whatsoever about where it had to go to get that information. That is because a living, evolving ecology has grown up around the Cloud bringing multiple clouds together seamlessly. This pervasive capability will allow our devices to maintain *context* on where we are, what we are doing, and what we are thinking, so that additional requests are better understood and more quickly acted upon. This means that any product or service in the future cannot assume it stands alone, but has to be able to deliver value based upon the *context* its consumer is currently in. This is closer to Kelly's concept of the Web as an *ecosystem* versus *organism.* It will be more a collection of organisms that support each other, in a shared environment that they all understand.
- *Everything as a Service:* Basically, there will be no more stand alone products. Cars call home and feed information to the Internet, so do medical devices. Our phones are jabbering constantly to the Web so that our any request can be instantly met. It used to be called the extended product, the physical product plus its manuals, its accessories, and so on. Now everything is an outcome delivering service that creates value for the consumer which may or may not involve a physical presence of some sort. Everything is a Service—think Zipcar's car sharing versus owning a car.

So why this sojourn around the world of the future of the Web? It's because "the Cloud" is *the* ecosystem upon which most business innovation will be based going forward. To understand the possibilities of innovation, one needs to understand the possibilities of the Cloud and how we have missed seeing these opportunities in the past. According to Peter Drucker, the general sources of innovations are different changes in industry structure, in market structure, in local and global demographics, in human perception, mood and meaning, and in the amount of already available scientific knowledge. All of this is fast becoming encapsulated inside the Cloud.

More and more buying and delivering value is taking place in the Cloud. When an innovative idea requires a better business model, or radically redesigns the delivery of value to focus on the customer, a real-world experimentation approach increases the chances of market success. Potentially innovative business models and customer experiences can be tested more quickly and even more broadly through the Cloud than through traditional market research methods. The Cloud will accelerate the diffusion of innovations throughout the economy.

If one wants to study fish, sea mammals, mollusks, and jelly fish, one has to understand the ocean, the environment they all exist in. To study business innovation in the 21st century, one has to understand the Cloud, that is, the environment that business and the economy will operate in.

Where Does Innovation Come From?

So where does innovation come from? One answer lies in the notion of *clusters.* American innovation isn't restricted to the research labs of the big companies or the famous universities. America innovates to the extent it does because there is a cluster of technologists who live in concentrated areas, e.g. Silicon Valley or Boston's Route 128, and frequent the same bars. As international software development and R&D clusters are moved to China, India, Russia and Korea, innovation will follow and increasingly become stateless.

In 2004, China began speaking of *technological nationalism*, meaning that the country wants make innovation an indigenous national asset, not an import. The Chinese don't just want to make what others innovate, they want to innovate and dominate global markets with their innovations.

But wait, what exactly is innovation?

A dictionary definition doesn't help much: 1: the introduction of something new, 2: a new idea, method, or device. Most often the tendency, when pondering what innovation is, is to think of a glitzy new product with bling, such as the iPad. But when we think of innovation as the Next Big Thing in business, we need to understand it in a business context that includes many variables besides product innovation.

Business innovation has multiple dimensions that interact to form a true breakthrough for competitive advantage. Further, the word "innovation" is problematic. Strictly speaking, an innovation is something *completely new,* but there is, practically speaking, no such thing as an unprecedented innovation in business or technology. Even in the world of science, true scientists will tell you of an invention credited to them as really being a product of "climbing on the shoulders of others"—their peers and predecessors.

With business innovation, it's usually about "connecting the dots" across the major types of business innovation to create something distinctive, something new, especially as perceived in the eyes of a company's customers. It's also about connecting the dots between unique business practices in other industries.

British business process consultant, Mark McGregor, describes how best practices can be drawn from several industries to create what he calls *next practices*, "What if you looked to brand-based companies such as Coca Cola for your ideas on marketing, what if you looked at someone like Amazon for your inspiration in building on-line shops for your products, and possibly someone like McKinsey as your inspiration for providing service? I am sure you will agree that a company that delivered products to the same quality as a pharmaceutical company and services to the standard of McKinsey, while being as smart at brand awareness as Coca Cola and as easy to buy from as Amazon—would cause more than a few ripples in its marketplace."

Returning to the idea that there are several major types of business innovation, let's set some context. At a high level, the simplest of business models is buy-make-sell. Thus companies have three key activities. They buy goods and services from suppliers. They add value to these inputs to make something of greater value than the sum of the parts, the inputs. Then they sell the good or service, hopefully at a margin that reflects the value added. That value is the value perceived by the customer, not just the sum of the costs of the parts that go into the good or service.

All of the buy-make-sell activities of a company consist of two types: *direct activities* that see the goods progress from acquiring the inputs through to producing and delivering the final product or service; and *indirect activities* that are essentially support activities, including facilities management, human resource administration, financial management, repairs, maintenance and so on. Direct activities ultimately touch the customer; indirect activities don't; hence the term, indirect.

All of these activities are objects of potential business innovation, taking us way beyond the simplistic notion that business innovation only equates to product innovation—which is usually an invention of some sort.

Everyone associates Thomas Edison with the light bulb. So strong is that association that many people actually assume that Edison actually invented the light bulb, whereas it was, in fact, invented by a man named Joseph Swann in Sunderland in the U.K.

What Edison and his team did was to perfect the light bulb and to create demand for such a product. It was Edison's "business innovations" that made money, not Swann's invention.

Xerox's Palo Alto Research Center (PARC) invented many of the technologies behind today's Macintosh PCs and iPads. But it was Apple Computer that made the money. In other words, product invention is but one type of "business innovation." A more complete list of business innovation categories (spelled out in the book, *Extreme Competition*) include:

- Operational innovation
- Organizational innovation
- Supply-side innovation
- Core-competency innovation
- Sell-side innovation
- Product and Service innovation.

Creative Destruction 2.0

Economist Joseph Schumpeter, who contributed greatly to the study of innovation, argued that industries must incessantly revolutionize the economic structure from within, that is innovate with better or more effective processes and products, such as the shift from the craft shop to factory. He famously asserted that "creative destruction is the essential fact about capitalism."

Schumpeter's theory of creative destruction provides a powerful explanation for the tectonic shifts that have taken place across the business and industrial landscape over the past two millennia. Out with the old and outmoded, in with the new, better and more adaptive. While it is hard for the companies, business models, and technologies that get creatively destroyed, there is a new net benefit to society overall.

The historian Thomas K. McCraw writes in his biography of Schumpeter, *Prophet of Innovation* (Belknap Press), "Schumpeter's signature legacy is his insight that innovation in the form of creative destruction is the driving force not only of capitalism but of material progress in general. Almost all businesses, no matter how strong they seem to be at a given moment, ultimately fail and almost always because they failed to innovate."

Mr. Schumpeter brilliantly realized that innovation—so often extolled as the purest expression of the human spirit—has a dark, violent, even nasty side. Every innovator, in short, makes a declaration of economic war. And every successful innovation is a destroyer of the closed system that came before it.

Barry Jaruzelski and Kevin Dehoff of Booz and Company commented in October 2009 that their firm's most recent Global 1000 Innovation Study indicated that despite the economic downturn companies are reluctant to reduce their innovation spending. Jaruzelski and Dehoff stated, "Innovation has become a core component of overall corporate strategy. Given the fierce nature of business competition in recent years, a reduction in innovation efforts would be akin to unilateral disarmament in wartime."

No one company is immune to the laws that govern systems, all types of systems. Companies that once revolutionized and dominated new industries—for example, Xerox in copiers or Polaroid in instant photography have seen their profits fall and their dominance vanish as rivals launched improved designs or cut manufacturing costs. But don't think that the innovators of digital photography that struck a death blow to the Polaroid can rest on their laurels, "It ain't over yet."

You may not recognize the name, Ren Ng, but you will. At age 31, Dr. Ng founded Lytro, the company that produced the focus free camera, revolutionizing digital photography. Remember cameras that would have to focus themselves before taking a snapshot? And how that could lose vital seconds, making a mockery of the term "point and shoot?" But now just about any amateur can become a professional sharp shooter with the game-changing Lytro camera.

Dr. Ren Ng, CEO, Lytro (www.lytro.com)

A related advance isn't just about taking pictures; it's about augmenting reality. Whereas *virtual reality* is about taking people into virtual digital worlds, Augmented Reality (AR) is about bringing the digital world into the real world where people actually live.

Augmented reality is a live, real-time, direct or indirect, view of a real-world environment whose elements are augmented by computer-generated input such as sound, video, graphics or GPS data. As a result, the technology functions by enhancing one's current perception of reality. By contrast, virtual reality replaces the real world with a simulated one.

AR is not a new concept. In fact, we've seen it in many different ways over the years, but we just might not have noticed. The yellow first-down lines sketched over a televised football game is an example of virtual graphics being superimposed upon a real-life situation. In 1990, Boeing researcher Tom Caudell first coined the term "augmented reality" to describe a digital display used by aircraft electricians that blended virtual graphics onto a physical reality. Some AR researchers say there's been hype for the idea of augmented reality since the 1930s. But the increasing number of AR smart-phone apps has created a whole new world for augmented reality, and for companies that learn to use it.

An AR smart-phone app generates a composite view for the user that is the combination of the real-world scene viewed by the user and a virtual scene generated by the computer that augments the scene with additional information. The virtual scene generated by the computer is designed to enhance the user's perception of the virtual world they are seeing or interacting with.

Today Augmented Reality is used in entertainment, military training, engineering design, robotics, manufacturing and other industries. Not all AR uses will be consumer oriented. For example the car maker, BMW, intends to use the technology to aid service technicians as demonstrated in this video.

youtube.com/watch?v=P9KPJlA5yds or tinyurl.com/37bpaj

While the worker in the BMW video used AR glasses, and Google is rumored to be developing heads-up display (HUD) sun glasses, we should mention the future of this sort of device, AR contact lenses.

And then there's sort of the reverse of augmented reality, that is, turning something virtual and making it into reality, *reality augmented*, if you will. 3D printing is an additive technology in which objects are built up in layers and usually over several hours. Whilst most 3D printers are currently used for prototyping and in pre-production mould making processes, the use of 3D printing to manufacture end-use parts is also now occurring. This is becoming known as direct digital manufacturing (DDM). For low-volume manufacturing DDM is more cost-effective and simpler than having to pay and wait for machining or tooling, with on-the-fly design changes and just-in-time inventory being possible. For example Klock Werks Kustom Cycles builds one-of-a-kind motorcycles using a Fortus 3D printer to directly digitally manufacture some of the required custom parts.

A patient specific titanium skull implant can be made with 3D printing using the Direct Metal Laser Sintering technology by EOS. Several thousand patients worldwide currently have 3D printed titanium implants in their hips, skulls or legs. In addition a total of 100,000 people have been helped with patient specific surgical guides that are 3D printed. Also, 10,000,000 people worldwide are currently wearing 3D printed hearing aids.

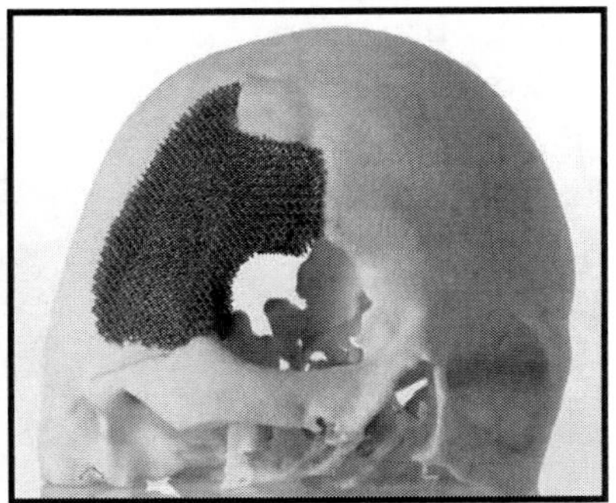

The future for 3D printing seems like science fiction, but it's not. Watch Wake Forest University's Anthony Atala discuss how *science fiction is becoming science fact* in regenerative medicine.

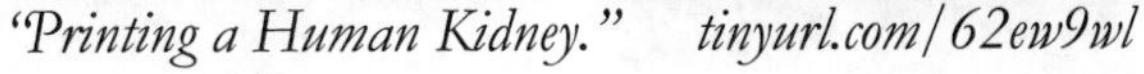

"Printing a Human Kidney." *tinyurl.com/62ew9wl*

According to *The Economist* in a February 10, 2011 article, "3D printing makes it as cheap to create single items as it is to produce thousands and thus undermines economies of scale. It may have as profound an impact on the world as the coming of the factory did. Just as nobody could have predicted the impact of the steam engine in 1750—or the printing press in 1450, or the transistor in 1950—it is impossible to foresee the long-term impact of 3D printing. But the technology is coming, and it is likely to disrupt every field it touches."

In his blog, *The Second Digital Revolution*,[2] British professor, Christopher Barnatt writes, "A decade ago, virtual reality was the next frontier. Daily journeys were predicted into computer-generated worlds in which any dream could become a reality. Work and play were both expected to be transformed as the human race joined computer data in cyberspace.

"In the early twenty-first century, virtual worlds are still far from common. Human beings continue to live, work and travel in physical places built from concrete, wood, metal and glass. Use of the Internet may have become a mainstream human activity. However, virtual reality, whilst increasingly advanced, has never gone mass market.

"The reason for the above state of affairs is that what was not predicted a decade ago was today's 'Second Digital Revolution.' This involves the *mass atomization* of digital content, in addition to *mass digitization.* What this means is that whereas a decade ago the drive was to push things and people into the computer realm of cyberspace, today far more effort is being directed into pulling digital content back into reality. Or in other words, no longer is the intent to build new worlds within computers, but rather to build new computing and communications devices into the real world. The Second Digital Revolution subsequently reflects an age in which an increasing number of computing-enabled devices are permitting the everyday development of ubiquitous computing, with Internet-access and other digital technology almost constantly available.

"*Enter Atomization.* The most valued digital products will in future be those that can easily be transformed into physical reality for mass human consumption. This is because most people will continue to value that which they can see, hear and touch more than the mental abstraction of cyberspace, and/or their immersion into any computer-generated virtual world. As a result, the most critical future economic process is likely to be that of atomization -- of transforming bits into atoms—as opposed to digitization, whereby physical things are encoded into binary. The complimentary yet opposing processes of digitization and atomization may be illustrated as follows:

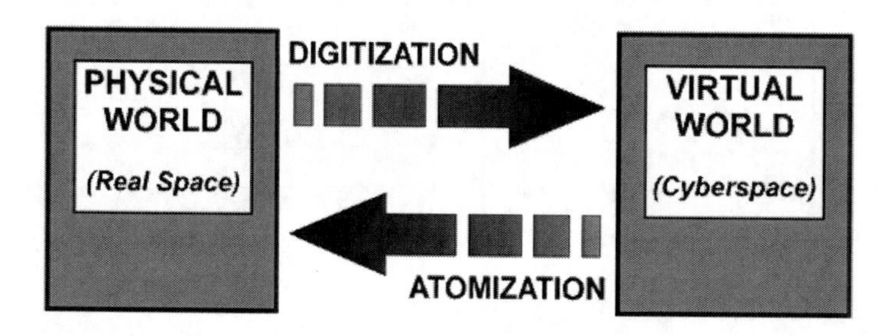

"The above figure allows us to clearly distinguish between and define the First and Second Digital Revolutions. Specifically, the First Digital Revolution involved only the top half of the diagram, and comprised the period of mass digitization that commenced around 1980, and during which time an increasing number of media, products and services were pushed into an electronic, binary format. In contrast the Second Digital Revolution involves all of the figure, and signifies the watershed that took place from around the year 2000 as mass atomization began to take hold, and during which time electronic, digital content has increasingly and routinely been 'pulled back' from cyberspace into the perceptibly-real world.

"Whilst personal computers are still clearly in mass use, so too are portable audio and video media players, personal digital assistants (PDAs) and ultramobile PCs, mobile phones with Internet multimedia access, and a variety of public forms of Internet access including in-store consoles and Microsoft's latest *surface computers* (microsoft.com/ surface). There are even now 3D printers capable of atomizing digital content not in the form of a video display or audio, but as real, solid physical objects. Hence, no longer is digital access something that requires human beings to conform to digital interfaces. Rather, the increasing divergence of digital access hardware now means that effectively digital technology is now being invented to meet human requirements. To put it another way, as the Second Digital Revolution really takes hold, the age of ubiquitous computing is very much starting to arrive."

What does all this mean for the status quo? For industry incumbents?

Walmart is a recent example of a company that has achieved a strong position in many markets through its use of innovative inventory-management, marketing, and personnel-management techniques, resulting in lower prices to compete with older or smaller companies in the offering of retail consumer products. Just as older behemoths perceived to be juggernauts by their contemporaries (e.g., Montgomery Ward and Woolworths) were eventually undone by nimbler and more innovative competitors, Walmart faces the same threat.

Amazon was never intent on becoming Earth's largest bookstore, it has always had the intent of becoming the world's largest "everything" store. Just as the cassette tape replaced the 8-track, only to be replaced in turn by the compact disc, itself being undercut by MP3 players, the seemingly dominant Walmart may well find itself an antiquated company of the past. This is the process of creative destruction in its technological manifestation.

Because a company is a "system," we can turn to the laws of systems theory. In animate thermodynamics we speak of entropy (a measure of the waste, chaos and disorder of a closed system) and its antonym, ectropy[1] (a measure of the tendency of a dynamical system to do useful work and grow more organized). In lay terms, the creative destruction effect of innovation means isolating the system that came before it, leading it to a state of entropy, while the redirected energy takes away the energy needed to maintain the former, now isolated, closed system.

Shumpeter's ideas of creative destruction go way back to 1939 and remind us anew that the pursuit of innovation is the moral equivalent of economic war. Companies that want to survive need to be building their arsenals. In Schumpeter's vision of capitalism, innovative entries by entrepreneurs was the force that sustained long-term economic growth, even as it destroyed the value of established companies and laborers that enjoyed some degree of monopoly power derived from previous technological, organizational, regulatory, and economic paradigms.

---

[1] When we eat, we take in ectropy from the food. The Second Law of Thermodynamics says that in a closed system, ectropy will decrease. An organism that is isolated from the outside world will die and deteriorate because its ectropy decreases. It needs ectropy coming from the environment to keep living.

So, while the process of creative destruction is by no means new, the Cloud enables Creative Destruction 2.0, creative destruction on steroids. With the potential of one shared computer, one shared information base and essentially one shared world, Creative Destruction 2.0 now proceeds at Internet time. Industry incumbents are now exposed to extreme competition from around the globe, not just within their once tidy national economies.

With goods now being global commodities, companies transition from brawn to brains as the ultimate capital resource. As we move to the Innovation Economy from the Knowledge Economy, knowledge management progresses to new knowledge creation. Where once it was "*an* innovation" that gave a company a reasonably long period of competitive advantage, it's now all about setting the pace of innovation (to wit, the smart phone and tablet industries). Innovation itself moves from R&D labs to open innovation. And once isolated knowledge-worker staff (e.g., R&D staff and those armies of MBAs) now gives way to every worker being a knowledge worker.

Nowhere is the pace of innovation and creative destruction more pronounced than in the tablet computer world. HP dumped its PC and tablet business (TouchPad) in 2011 (and then changed its mind within months) while Amazon announced a Kindle tablet.

Let's take a look at IT analyst firm Gartner to help us demystify innovation ...

Past View	Emerging View
▪ R&D and technology ▪ Breakthroughs ▪ Led by particular group ▪ Specialized teams and initiatives ▪ "Flash of genius" ▪ Benefits difficult to define and capture ▪ Significant investment ▪ Significant time to achieve benefits	▪ Can happen across all functions ▪ Continuous improvement *and* breakthroughs ▪ Organization-wide ▪ Ongoing process ▪ Product of a process ▪ Can be formalized in outsourcing relationships ▪ Requires little investment ▪ Benefits can be immediate

Source: Gartner

Creative Destruction 3.0

During the writing of this book, or more precisely on October 31, 2011 (on Halloween!), the world population cracked the 7 billion mark. In the year 1 AD, according to estimates from the U.S. Census Bureau, the world's total population was between 170 to 400 million. If mankind survives, the world's population in 2050 is estimated to be over 9 billion, the equivalent of adding two new Chinas.

"It's Time To Leave The Planet."—Prof. Stephen Hawking

"The Times They are a-Changin."—Bob Dylan

The opening up of new markets and the organizational development from the craft shop and factory to such concerns as US Steel illustrate the process of industrial mutation that incessantly revolutionizes the economic structure from within, incessantly destroying the old one, incessantly creating a new one.
—Joseph Schumpeter, *The Process of Creative Destruction*, 1942

Again, however, from destruction a new spirit of creation arises; the scarcity of wood and the needs of everyday life... forced the discovery or invention of substitutes for wood, forced the use of coal for heating, forced the invention of coke for the production of iron.
—Werner Sombart, *Krieg und Kapitalismus*, 1913

In short, industrialism is over.
—Paul Hawken

And I am a beneficiary of the promise of America. But today, I am very concerned that at times I do not recognize the America that I love.
—Howard Schultz, CEO, Starbucks

Millions of consumers and citizens are already convinced of a fact that many corporate chieftains are still reluctant to admit: the legacy model of economic production that has driven the "modern" economy over the last hundred years is on its last legs. Like a piece of clapped out engine, it's held together with bailing wire and duct tape, frequently breaks down and befouls the air with noxious fumes.
—Professor Gary Hamel

The conservation of natural resources is the fundamental problem. Unless we solve that problem it will avail us little to solve all others.
—Franklin D. Roosevelt

Indeed, we could write a whole book about the huge changes we are experiencing. But that would take us out of bounds for this book. We don't want to turn this book into a political discourse, we simply want to explore how today's political discourse on economics, however dysfunctional, is changing the landscape of business. What we want to point to is that …

> …the Great Recession of 2007+
> isn't about another *business cycle*, it's a *reset,*
> as big as the reset of the Industrial Economy
> that swept away the Agricultural Economy.

To keep the book from going too far out of its bounds, we've posted an exclusive 49-page PDF document for readers that are interested in more of the macro-level changes going on:

www.mkpress.com/CreativeDestruction3-0.pdf
or tinyurl.com/7vaz9an

Here's a quick guide to who and what is included:

- A primer from professor Richard Wolf, *Capitalism Hits the Fan.*
- Comedy Central's Jon Stewart on the *Fear Factory.*
- Aronica and Ramdoo's book, *The World is Flat?,* and "America's Former Middle Class."
- Economist Jeremy Rifkin on the *Third Industrial Revolution.*
- Paul Hawken, the entrepreneur behind the Smith & Hawken gardening supplies empire, on *Natural Capitalism.*
- Malcom Gladwell on the all-time disparity of the middle class income and that of corporate executives.
- Rachael Botsman on *What's Mine is Yours: The Rise of Collaborative Consumption.*
- Umair Haque, author of *The Capitalist Manifesto*, on an intellectual rebooting of the capitalist operating system.
- Harvard's Michael Porter on his thrust into creating *"shared value."*
- Roger Martin, Dean, Rotman School of Management, on "The Age of Customer Capitalism."
- Virgin group CEO Richard Branson on treating charity like a business.
- Steve Katsaros on getting his solar light bulb, "Nokero," to the 1.4 billion people around the world who don't have access to an electrical grid.

- Nobel laureate and author of *The Next Convergence*, Michael Spence on the prospects of the five billion people who live in developing countries.
- "Dr. Doom," economist Nouriel Roubini for additional insights into the new normal.
- Professor Gary Hamel on capitalism as it's the worst sort of system—except for all the others. His 2012 book is *What Matters Now.*
- Economist Fredrick Heyak's 1946 observations on "The Use of Knowledge in Society,"
- Sramana Mitra, a Silicon Valley consultant, on the One Million by One Million initiative to help a million entrepreneurs globally.
- Professor Nassim Nicholas Taleb on an economic life closer to our biological environment
- Berkeley economist and author of *Aftershock: The Next Economy and America's Future*, Robert Reich, on the anxiety gripping the middle class, and their inability to go on buying enough to keep the economy going

And so it seems that Creative Destruction 3.0 is well underway. Globalization is far more than multinational corporations going to the ends of the earth for cheap labor. From China's CommuCapitalism to Mercedes Benz CapitalCommunism, something is going on.

"Some colleagues still think that car-sharing borders on communism," Mercedes-Benz Chairman of the Board of Management Dieter Zetsche said onstage at the Consumer Electronics Show in 2012, speaking about Mercedes' new CarTogether initiative. "But if that's the case, viva la revolucion!"

To be sure, a luxury-car maker like Mercedes is not actually promoting communism. But during his CES talk, Zetsche pushed hard on a vision that the company has for a *greener future* that allows drivers to reduce emissions by using connected and social technology to easily find compatible passengers to share rides with.

But be sure that Zetsche's talk signals a huge transformation going on that some call Capitalism 2.0 while others are going so far as Capitalism 3.0.

> Huge transformations have been going on
> as globalization resets the economic playing field,
> and the very game itself.

Deng Xiaoping gave us a perhaps one of the greatest examples by allowing China to become CommuCapitalists, and become a world leader in Capitalism 1.0. Are we on the verge of a similar-scale transformation to CapitalCommunists?

The debates are raging.

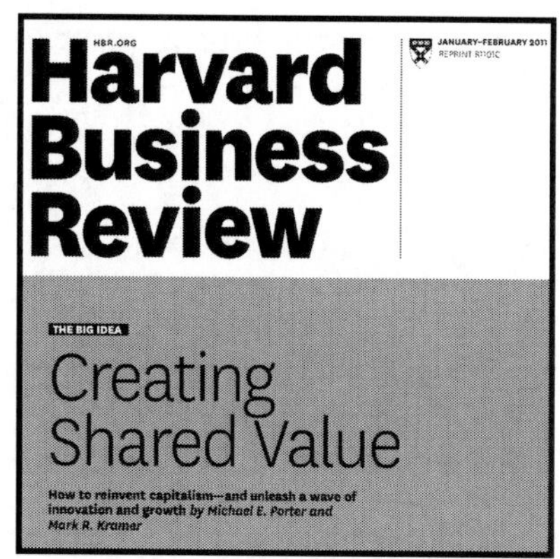

Watch: tinyurl.com/87ew27e

Watch Malcom Gladwell on Income Inequality *tinyurl.com/4xeslcs*

New Unanswered Questions Need to be Asked. Here are a few unanswered questions from Martin Ford's book, *The Lights in the Tunnel: Automation, Accelerating Technology and the Economy of the Future:*

- "How will job automation impact the economy?
- How will the offshore outsourcing trend evolve?
- What impact will technologies such as robotics and artificial intelligence have on job markets?
- Did technology play a significant role in the 2007 subprime meltdown and the subsequent global financial crisis and recession?
- Globalization. Collaboration. Telecommuting. Are these the forces that will shape the workplaces of the future? Or is there something bigger lurking?
- How fast can we expect technological change to occur in the coming years and decades?
- Which jobs and industries are likely to be most vulnerable to automation and outsourcing?
- Machine and computer automation will primarily impact low skilled and low paid workers. True or false?

- Will advancing technology always make society as a whole more wealthy? Or could it someday cause a severe economic depression?
- What are the implications of advancing automation technology for developing nations such as China?
- Will a college education continue to be a good bet?
- Recent economic data suggests that, in United States, we are seeing increasing income inequality and a dwindling middle class. How will this trend play out in the future?
- What will be the economic impact of truly advanced technologies, such as nanotechnology?
- Retail positions at Walmart and other chain stores have become the jobs of last resort for many workers. Will robots and other forms of machine automation someday threaten these jobs? If so, what alternatives will the economy create for these workers?"

We explore aspects of the current Services Economy in Chapter 15. Of note, 80% of employment in the U.S. is now in services. Two hundred years ago, 90+% of U.S. employment was in agriculture. And now science is just beginning to be applied to services, while science has already been applied to agriculture (now just 2% of employment) and manufacturing (now just 12% of employment). "It's all about the substitution of information for labor to solve these pressing human problems," says IBM's Robert Morris. "Whoever figures out that substitution will triumph."[3]

If science is successfully applied to services
—and the rush is on—
will we see the advent of the jobless society?

What good is it making products if there are few consumers, thanks to the "jobless economy?" Consumers need incomes. Even conservative free-market economist, Milton Friedman, once proposed a Negative Income Tax (NIT), a progressive income tax system where people earning below a certain amount receive supplemental pay from the government instead of paying taxes to the government.

So, what's a business to do amidst such high levels of macro-level uncertainty?

According to the late Peter Drucker, "Trying to predict the future is like trying to drive down a country road at night with no lights while looking out the back window."

As we are encountering a world of the unknowable, sense and respond *agility* is the key to surviving and thriving in the 21st century.

Agility—that elusive word of business—
is the cornerstone for surviving and thriving in the 21st century.

But the idea isn't new at all as we can learn from Charles Darwin, "It is not the strongest of the species that survives, nor the most intelligent, but the one most responsive to change."

We might add "and most rapidly takes advantages of the resources that appear." Again, the Cloud makes all of the technologies, knowledge and capabilities we have been discussing "instantly" available to everyone—and the most innovative will apply them rapidly.

Because we live in an age of hyper-connectivity driving an unprecedented global economy, where and how we conduct business is of prime importance. That "where" is in the Cloud, and the "how" is multi-company, socially-driven, business process management, for BPM is how works gets done.

And, returning to the main thought opening this book, *doing the work is all* in 21st century business competition and maintaining the relevance of capitalism. Let's have no demi-God to bail us out of the current global crises as the down-trodden Germans once thought one could do in the 1930s. Let's instead embrace business innovation in the Cloud using pilot John Boyd's energy maneuverability theory and OODA loops presented in Chapter 5 to gain the *agility* we need in a rapidly changing world.

Tenets of Innovation

Permanent Innovation: The Essential Guide to the Strategies, Principles and Practices of Successful Innovators, is a free downloadable book that explores the essential strategies, principles, and practices of permanent innovation to help your company to become a permanent innovator, and perhaps even a leader. In it several principles for innovation are discussed and abstracted here from the overview. [4]

- *Innovation is essential to survival, and all innovation is strategic.* As innovation is literally how organizations create their own futures, innovation as a process and an organizational priority cannot be separated from the development and implementation of strategy. Hence, the development of a productive innovation capability is one of the most important strategic priorities for any organization. At the same time, all innovation must be guided by strategic priorities and intentions.
- *There are many types of innovation: e.g., incremental, breakthrough or disruptive made manifest in products and technologies, new business models, and new ventures.* Taken together, all of our innovation initiatives constitute a portfolio. As we design the portfolio we'll have to decide how much effort and investment to allocate to each type. Each requires its own specific set of processes, tools, and teams, who will be engaged in the search for the future, which is what the search for innovation is all about.
- *The longer we wait to begin innovating, the worse things will get.* Companies that procrastinate usually pay a heavy price in the form of lost market share and lost profits, and ultimately the lack of innovation can significantly diminish their future prospects. The competition isn't waiting, and you

shouldn't either. Create the action plan now, and start implementing it now!

- *Innovation is a social art; it happens when people interact with one another.* People are the core of any innovation process. Their insights, concerns, and desires shape the pursuit of new ideas and the countless decisions to be made in the process of transforming those ideas into value. Consequently, managing innovation is largely a process of managing people, and also managing the principles and practices according to which their work is organized. This requires a great deal of thought, planning, and preparation.
- *Innovation without methodology is just luck.* There are lots of creative people in a company, and given half a chance they'll probably create some great innovations. But if we rely on their random efforts then we're risking our future success on chance, and that's not enough. We have to develop and apply methodologies, the right methodologies, to make the shift from luck to consistency, predictability, and sustainability. Without the right innovation methodology we're risking far too much—we're risking the future.
- *Strategic innovation viewpoints are critical to success.* You can't rely just on the innovation efforts of top managers, nor of people in the field, nor of what only insiders can create. The complete innovation methodology has to leverage all four viewpoints: Top down, Bottom-up, Outside-in, and Peer-to-peer.
- *Great innovations begin with great ideas; to find them, identify unknown and unmet needs.* There are many different kinds of needs. Among the most significant for innovators are the ones that no one has recognized, for these offer the potential to create breakthroughs that bring significant added value and competitive advantage. So how to find them? There are dozens of tools we'll explore later in the book that we can apply to come up with new ideas.
- *Ready, Aim, Aim, Aim, Fire.* Yes, it's a cliché. But it's also true. Effective innovation requires very careful targeting. Why? Because there are so many possibilities to chase that we have to make sure we're going after the right ones. Besides which, innovation can be expensive both in terms of cash and time, and good aiming enables you to use your resources wisely. At the same time remember our previous warning of the risks of target fixation!
- *Prototype rapidly to accelerate learning.* The goal of any innovation process is to come up with the best ideas and get them into action as quickly as possible. Thus, the innovation process is a learning process, and learning faster has enormous advantages. Among the methods for learning that we can choose, prototyping is one of the most valuable because it so effectively condenses the learning process. Rapid prototyping is thus central to most forms of effective innovation methodologies.
- *There is no innovation without leadership.* Companies are amazing expressions of human society. The organizing of thousands of people to create and deliver products and services around the world to thousands or millions of customers is a remarkable thing. But the ability to do this brings some unique challenges. In particular, the impact of the organizational hierarchy has tremen-

dous influence on the culture of any company, on its ways of working and the results it achieves. Thus, top managers can be powerful champions of innovation, or dark clouds of suppression. It's up to leaders to ensure that their words and actions support and enhance innovation efforts and methods, and that at the same time they work diligently to eliminate the many obstacles that otherwise impede or even crush both creativity and innovation.

- *It is the established companies that are having the bigger struggle with innovation.* Any company, no matter its size, must redefine the way in which it can be of service to its customers. Companies must reengineer their processes to be more adaptive. Companies must improve and enhance, and on occasion reinvent, their product and service lines—and do it all as innovatively as they can, with the goal of creating outcomes that the competition can't match.

Innovation's Relationship to Creativity

Innovation involves creativity, but does not equal creativity. "All innovation begins with creative ideas . . . We define innovation as the successful implementation of creative ideas within an organization. In this view, creativity by individuals and teams is but a starting point for innovation (Amabile, 1996)."

So if we build on the premise that all innovation begins with creative ideas, then, should we not be asking what are the sources of creativity and what drives creativity? Based on the research of Teresa Amabile, PhD in Psychology and Head of the Entrepreneurial Management Unit at Harvard Business School, we see that creativity stems from the convergence of Expertise (Knowledge), Creative Thinking skills, and Motivation:

- *Expertise (Knowledge)*—all the relevant information that an individual brings to bear on a problem or subject. This can be a long-time focus in a specific area or the ability to combine previously disparate elements in new ways. Individuals who have a broad focus with varied interests across multiple disciplines also fit well into this category.
- *Creative Thinking*—regarding creative thinking, one aspect centers around the environment that facilitates the freedom to think creatively and the individual's comfort level in expressing different ideas.
- *Motivation*—more important than cognitive abilities is intrinsic motivation toward the positive impact on creativity. "People are most creative when they feel motivated by the interest, satisfaction and the challenge of the work itself and not by external pressures (extrinsic motivation)."

Ultimately, by first understanding how creativity works and how we can facilitate creativity, we can then create the atmosphere and environment that will nurture creative ideas to bring forth innovation.

Barry Jaruzelski, a Booz Allen Hamilton Vice President, has commented about his firm's

Global 1000 Innovation Study, "We examined an extensive database of patents and found that if companies spent more money on R&D, they did get more patents, but the level of patent activity did not correlate with corporate success as measured in terms of growth, profitability, and shareholder return (Bernstein, 2008)." Jaruzelski also indicated in the same article that innovation and new technology are not equivalent. Specifically, he commented, "New technology may be a technical advance, but it's not a real business innovation unless it drives significant new revenue streams."

Consider the iPod. The true innovation was iTunes and the one-stop-shopping, ease-of-use business model. There was no significant technical advance. MP3 players had been out there for a few years, so had online purchasing and downloading of music. The real innovation was getting all those music catalogs under one roof and making them easily accessible by implementing established technologies in a simple, coherent way. Those were business model innovations that transformed the digital music industry.

Three Core Types of Innovation

A lot that has been written on the topic of innovation and people have categorized the types of innovation in various ways over the years. In our view, there are simply three core types of innovation: Sustaining, Breakthrough and Disruptive innovation.

Sustaining Innovation refers to the continual, incremental improvements in existing goods, services, and processes to increase value. This type of innovation focuses on incremental performance and productivity enhancements of existing products and services that come from improved materials, technologies, sources of capital, streamlined process flows, better worker training, improved supply chains, and many other ways. This type of innovation is generally consumer-driven. Six Sigma methods, theory of constraints analysis and process re-engineering are all approaches that result in sustaining innovation which can generally be predicted in terms of costs to create, time to manifest and resulting value generated. As a consequence the market generally expects and anticipates this form of innovation and factors it into its decision criteria and purchasing patterns.

Breakthrough Innovation is the introduction of *new usage patterns* that leverage existing technology, products and services in novel combinations, using current capabilities, that are applied cleverly to create a new value proposition within an existing business process or value delivery system. Also, breakthrough innovations can be totally new business models or processes, again leveraging existing technology, products and services. Or they can be the introduction of new technologies, products and services within existing business models and processes generating totally new outcomes or value propositions. Breakthrough innovations could be considered a discontinuous form of a sustaining innovation in that they create a new usage pattern that is unexpected. They are more predictable than a disruptive innovation in that they can be somewhat anticipated.

Disruptive Innovation is the introduction of radically different, unexpected products, processes, or services into the market that deeply impacts lifestyles and purchasing patterns and decision criteria and can create an entirely new segment of consumers. This type of innovation can cause significant leaps in value delivered to customers, although historically disruptive innovation has required longer adoption periods because they generally require consumers to change current behavior to something very new and different. Disruptive innovation is usually design-driven, involves the application of new technologies, often times invokes a new business model and may result in the cannibalization of existing sales as current products and services become superseded by the new innovative product or service. Often a disruptive innovation is the result of combining multiple breakthrough innovations. For example, the iPod was a breakthrough innovation on MP3 players (form, ease of use, etc.). The service iTunes was a breakthrough business model for the delivery of content. The two combined to create a content consumption disruptive innovation with technology, process and business model implications.

Disruptive innovation does not mean the elimination of previous technology, processes or business models, but does force change in them (bias ply tires 40 years ago and today after the introduction of radial tires). Disruptive innovation can also bring with it unintended consequences for industry incumbents, such as the rapid movement to Internet telephony has done to analog telecommunications services.

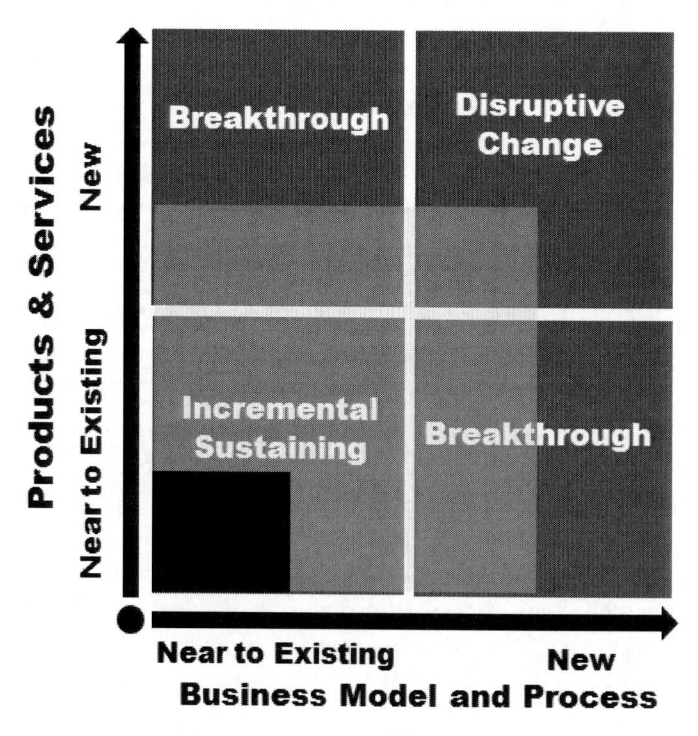

It is important that regardless of the type of innovation under consideration, *the degree* with which the innovation changes the way things are done has implications on manageability, value and the continuity of business. At some point innovations move from being simply pragmatic (better, faster, cheaper ways of doing what we always have done, as represented by the darker box in the diagram above) to being transformative (we don't do it that way anymore, as represented by the lighter box).

S-Curves and Innovation

Innovation is actually a decision making process, both by the developer of the innovation and the consumer of the innovation. Innovation consumption is best represented by the famous S curve, or as it is technically called—"the S-shaped diffusion curve." This curve was first developed by Gabriel Tarde in 1903 wherein he described the innovation consumption process as five steps:

1. First knowledge
2. Forming an attitude
3. A decision to adopt or reject
4. Implementation and use
5. Confirmation of the decision

No matter how formal or informal the innovation initiative, using a process to transform latent ideas into value-creating innovations is recommended. There are vast numbers of innovation processes out there, but all encompass the following steps, which we elaborate on elsewhere in the book: 1) Scope, 2) Capture ideas, 3) Evaluate and select, 4) Develop, 5) Implement, 6) Champion.

Part of innovation, *championing*, is influencing the consumption of the innovation. Manufacturing Consent is a method of championing innovations in advance, and we'll explore this notion further in the Chapter 8, Innovation and Social Networks.

Once innovation occurs, innovations may be spread from the innovator to other individuals and groups. In the early stage of a particular innovation, growth is relatively slow as the new product establishes itself. At some point customers begin to demand the innovation and the product growth increases more rapidly. Then, incremental innovations or changes to the product allow growth to continue. Toward the end of its life cycle, growth slows and may even begin to decline. In the later stages, no amount of new investment in that product will yield the initial rate of return. Peter Denning and Bob Dunham contend in their book, *The Innovators Way*, that focusing on ease of adoption is more important for successful innovation than idea generation and development.

The s-curve derives from an assumption that new products are likely to have a *product lifecycle:* a start-up phase, a rapid increase in revenue and eventual decline. As a matter of fact, a great majority of innovations never get off the bottom of the curve, and never produce desired returns.

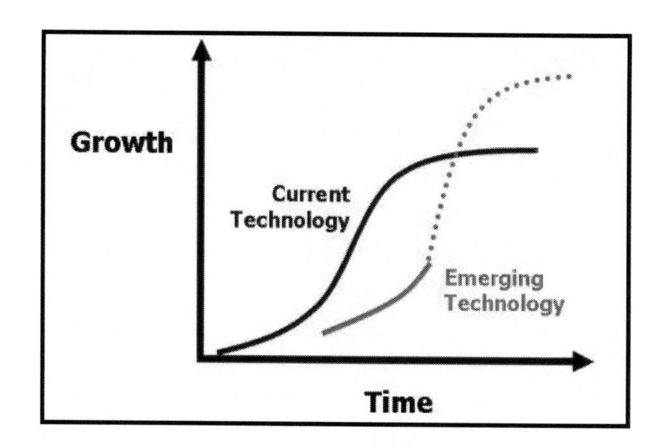

Successive s-curves will come along to replace older ones and continue to drive growth upward. In the figure below the first curve shows a current technology. The second shows an emerging technology that initially yields lower growth but will eventually overtake current technology and lead to even greater levels of growth. The length of life will depend on many factors.

Why Innovate?

Innovation is the key to growth of any business. There are three ways to grow a business:

- *Grow with the Market:* Growing with an expanding market is only possible by relying on past innovation. Once established as a viable player in a growing market, it takes comparatively less effort to sustain that position. This is the case of "the rising tide lifts all boats." Sustaining Innovation is key to growing with the market.
- *Increase Market Share:* To increase market share there must be some level of market differentiation. While continual and creative marketing can improve market share by itself, typically it's due to innovative approaches in features, functionality, quality, consumer choice and bundling, or even new business models that distinguish one product or service from a competitor's. Increasing market share is generally a result of programmatic Sustaining Innovation and the occasional Breakthrough Innovation.
- *Create or Expand into New Markets:* In order to create or expand into new markets, especially in technology, programmatic Breakthrough Innovation and the occasional Disruptive Innovation is required.
- Lastly, *innovation drives margin*: Sustaining innovation generally reduces "transaction costs." Breakthrough innovations can both reduce costs and increase customer value. Disruptive innovations, by definition, temporarily allow effectively what economists call monopoly profits for a period of time—punctuated equilibrium.

Takeaway

Innovation is not a marketing term! Innovation must be real. Everyone assigns a different meaning to the word and marketing organizations are destroying the word. It loses its meaning because there is not a common definition. Unfortunately, the same is true for "the Cloud," only this time the IT department is as guilty as the marketing department.

There are five elements to an innovation. It must be a new idea. It doesn't have to be brand new. It could be an idea borrowed from another place and given a new implementation to achieve a new, different outcome.

It has to be forward thinking. There are exceptions to the rule when an idea from the past is recovered and applied in an entirely new context. But, in general an innovation is created in anticipation of a future state of the environment (even though the problem it solves or pain it eases might exist in the present).

An innovation has to be feasible. The greatest idea in the world has no value if it cannot be made manifest. Likewise, it has to be viable – not just a one hit wonder, it has to be sustainable and provide extended use over time and markets. It is also not viable if it is an idea before its time and will fail the value test as well. It has to become self-supporting at some point in time. Think of the glacially slow adoption of PDAs compared to smart phones. (Essay test: What is different about smart phones versus PDAs?)

Lastly, it has to be valuable. Someone has to be willing to give up value (e.g., money) in order to acquire the value the innovation supplies or the outcomes it creates—otherwise why bother.

References

[1] http://www.slideshare.net/cvgallo/7-innovation-secrets-of-steve-jobs

[2] http://www.explainingthefuture.com/sdr.html

[3] money.cnn.com/galleries/2010/technology/1007/gallery.smartest_people_ tech.fortune/41.html

[4] http://www.permanentinnovation.com

Part 2.
Getting There: Ingredients for Executing on Innovation

Some promote the theory that innovation is a fortunate series of events that cannot or should not be managed. Others think of innovation as a "eureka" moment where great ideas spring fully formed from minds of eccentric, creative people.

Many organizations centralize responsibility for innovation in a small team, incubator or a research and development (R&D) group. Such organizational structures have advantages. Such teams usually work in isolation and these approaches can speed innovations to market, separate innovators from day-to-day operational issues and bring together key experts to collaborate.

Despite these advantages, innovation is too important to be left to the R&D organization alone. The location of valuable new ideas is migrating away from the center toward the *edge*. As we move into a service and knowledge economy, the number of people with useful information that can lead to good ideas multiplies exponentially.

The key is to maintain a multi-dimensional and divergent view of all the trends and ideas in as many domains as possible. Then qualitatively translate those into the value systems and requirements that will be used for making decisions. Likewise, you must stay rooted in the constraints that are in place, because changing status quo is hard.

While there is no serial, cookie-cutter approach for excelling at innovation, there are several *work ingredients* that are needed. Part 2 of this book documents those ingredients that, taken all together, provide the winning recipes for doing the work of business innovation.

None of these work ingredients may be new to your organization. An ingredient may have been pursued individually in your organization. However, when one of these ingredients is pursued on its own without the others, it tends to be unsustainable because it conflicts with the goals, attitudes and practices of the existing organization. The ingredients are interdependent. When the ingredients are accounted for simultaneously, the result is sustainable change that is radically more productive for the organization, more congenial to innovation, and more satisfying both for those doing the work of business innovation and those for whom the work is done.

5. Business Agility and the Dogfight in the Cloud

"Once there is seeing, there must be acting.
Otherwise, what is the use of seeing?"
— Thich Nhat Hahn

Did you ever wonder how a smaller, slower, shorter-range and lower-altitude jet fighter could beat its far more endowed enemy aircraft in a supersonic dogfight?

Well, we could turn to the "Father of the F16," a cigar-smoking, cocky, foul-mouthed guy for answers. Over the years many have written about this fellow, but his lessons are being revisited by smart companies in light of the Great Recession and the New Normal (a meteor hit the economy, and we've been living in the crater ever since). Some might even say this is old stuff. But sometimes what's old is new again. Following is a recap of some of the current writings.

The late John R. Boyd, a U.S. Air Force fighter pilot trainer, issued a standing challenge to all comers. Starting from a position of disadvantage, he'd have his jet on their tail within 40 seconds, or he'd pay out $40. Stories have it that he never lost. Boyd's maverick lifestyle and unfailing ability to win any dogfight in 40 seconds or less earned him his nickname, "40-Second Boyd." According to his biographer, Robert Coram, Boyd was also known at different points of his career as "The Mad Major" for the intensity of his passions, as "Genghis John" for his confrontational style of interpersonal discussion, and as the "Ghetto Colonel" for his spartan lifestyle. Boyd rarely met a general he couldn't offend.

In a dogfight, the F-16 seems to ignore the laws of physical science. Its design allows extreme maneuvers. It turns energy on and off in a second, and despite its light weight, it can withstand nine times the force of gravity, which enables some serious twisting and rolling. The plane is unbelievably agile.

The F-16 allows its pilot to *outmaneuver* the other guy, just as a company would like to be able to outmaneuver its competitors who are bigger, stronger and faster. Likewise, by being able to sense and respond to the changing competitive environment with great speed, the agile business can confuse its competitors, and by the time they have figured out what happened, the agile business again goes on the offensive in the very environment it had just created.

Though Boyd certainly had defensive strategies, he counted mostly on overwhelming offensive tactics. Boyd is credited for largely developing the strategy for the invasion of Iraq in the first Gulf War. In 1981 Boyd had presented his briefing, *Patterns of Conflict,* to Richard Cheney, then a member of Congress. In 1990 the U.S. Secretary of Defense, Cheney, called him back to work on the plans for Operation Desert Storm. Boyd had substantial influence on the ultimate "end around" design of the plan. In a letter to the editor of *Inside the Pentagon*, former Commandant of the Marine Corps General Charles C. Krulak is quoted as saying, "The Iraqi army collapsed morally and intellectually under the onslaught of American and Coalition forces. John Boyd was an architect of that victory as surely as if he'd commanded a fighter wing or a maneuver division in the desert." One amazing aspect of Boyd's offensive strategies was the sight of American columns running up both banks of the Euphrates with little apparent concern for securing their rear or maintaining a supply route. It was an *offense-first* strategy.

During the early 1960s, Boyd, together with Thomas Christie, a civilian mathematician, created the Energy-Maneuverability theory of aerial combat. E-M theory is a mathematical equation that calculated an aircraft's ability to change direction, altitude or speed under specified conditions, and allowed systematic comparison between different aircraft for the first time. In the 1970s, Boyd applied his understanding of energy maneuverability to help design the F-16, bringing together everything he knew about competition. Then he focused on something even grander, a unifying theory of this thing we call *agility*.

Boyd's key agility concept was that of the *decision cycle* or OODA Loop (Observe, Orient,

Decide, Act), the process by which either an individual or an organization reacts to an event in real-time, or at least in time enough to make a difference. According to this idea, the key to victory is to be able to create situations wherein one can make good decisions *more quickly* than the opponent. The construct was developed out of Boyd's earlier E-M theory and his observations on air combat between MiGs and U.S. Sabre jets in the Korean war. The chief designer of the F-16, Harry Hillaker said of the OODA theory, "Time is the dominant parameter. The pilot who goes through the OODA cycle in the shortest time prevails because his opponent is caught responding to situations that have already changed."

Boyd hypothesized that all intelligent organisms and organizations undergo a continuous cycle of interaction with their environments. Boyd broke this cycle down to four interrelated and overlapping processes through which one cycles continuously:

- *Observe:* the lightening-quick collection of *relevant* information about your current environment by means of the *senses* rather than drawn-out data analysis that leads to "analysis paralysis." Picking the "right data" to focus on is essential in order to avoid information overload.
- *Orient:* the analysis and synthesis of information to form one's current mental perspective. This is the most important step because this is where information is turned into "situational awareness" not unlike what's needed in a 3D chess game.
- *Decide:* the determination of a course of action based on one's current mental perspective. Again, avoid analysis paralysis, as a dogfight happens in real time.
- *Act:* the physical carrying-out of the decisions. Of course, while your action is taking place, it will change the situation, favorably or unfavorably. Hence the loop repeats, again and again in real time as nothing is standing still.

In the course of a dogfight, a pilot will go through many OODA Loops, with the number depending on the complexity of the situation. The pilot that can go through the repeating loops the fastest gains a distinct advantage, especially if she can act in ways that confuse the enemy combatant. In other words, the winning pilot is not just dealing with her own loops, the pilot is funneling the chaos of the dogfight to overload her opponent's OODA Loops. That is, she is operating *inside* an adversary's OODA loop to, as Boyd wrote, "... make us appear ambiguous, and thereby generate confusion and disorder." If the pilot is ahead in what Boyd calls *fast transients,* the opponent slowly loses touch with reality, and gets shot down. Think "blitzkrieg."

Want to see a F-16 Dogfight?

www.youtube.com/watch?v=INb-421E-mo
or tinyurl.com/yz9nvj7

As Keith Hammonds cited in *Fast Company*, "Boyd wrote, 'The winner collapses his adversary's ability to carry on.' You win the competition by destroying your opponent's frame of reference. Boyd saw isolation as a critical strategic device—in effect, the opposite of the information-rich environment that pilots (or companies) need in order to operate effectively. In isolation, he argued, a competitor had no hope of observing and adapting to a changing environment. Isolating your enemy, Boyd saw, could become a powerful tool to make his OODA loop inoperable, cutting off the flow of information both in and out of the organization." [1] The effective use of social networks is one means that can be used for isolating the competition in business (see Chapter 8).

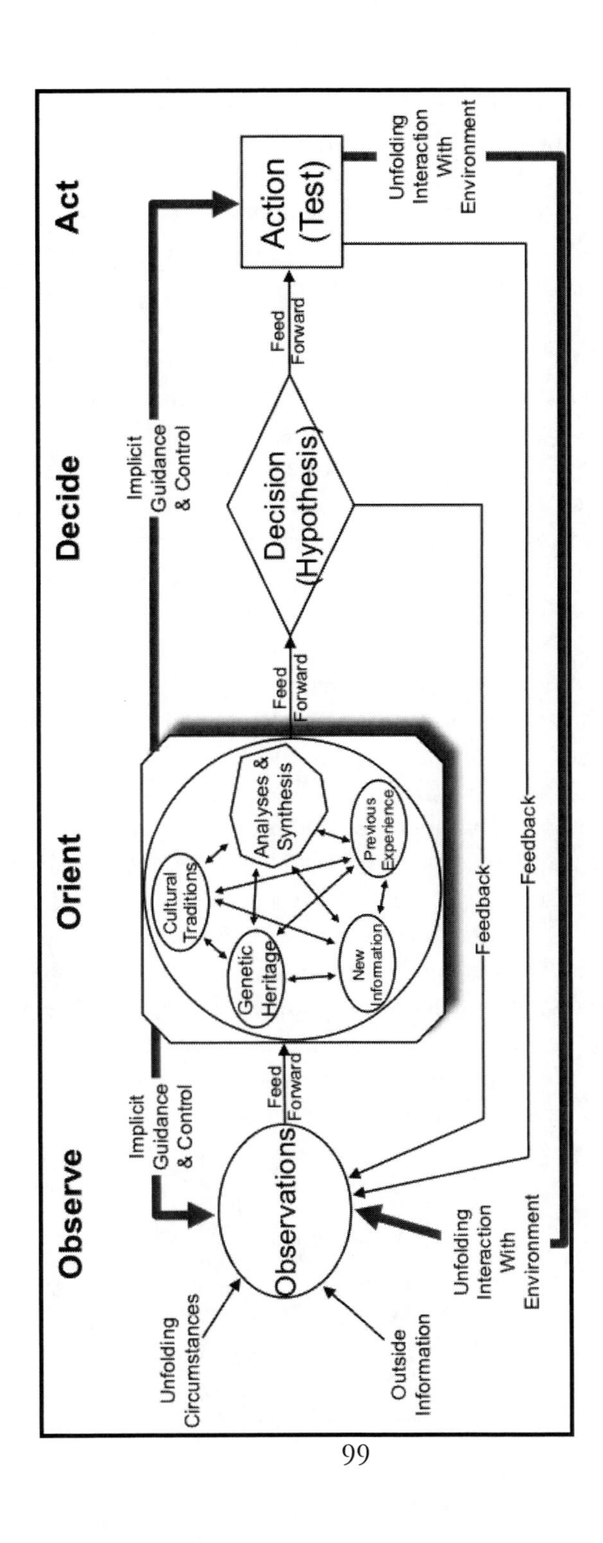
Observe
Orient
Decide
Act
Unfolding Circumstances
Outside Information
Observations
Implicit Guidance & Control
Feed Forward
Unfolding Interaction With Environment
Cultural Traditions
Genetic Heritage
Analyses & Synthesis
New Information
Previous Experience
Feed Forward
Feedback
Feedback
Implicit Guidance & Control
Decision (Hypothesis)
Feed Forward
Action (Test)
Unfolding Interaction With Environment

What does all this dogfight stuff have to do with business? Robert Greene writes in *The 33 Strategies of War* (tinyurl.com/664lwaf), "This seemed to me the perfect metaphor for what we are all going through right now in the 21st century. *Changes are occurring too fast for any of us to really process them in the traditional manner.* Our strategies tend to be rooted in the past. Our businesses operate on models from the 60s and 70s. The changes going on can easily give us the feeling that we are not really in control of events. The standard response in such situations is to try to control too much, in which case everything will tend to fall apart as we fall behind. Those who try to control too much lose contact with reality, react emotionally to surprises. Or to let go, an equally disastrous mindset. What we are going through requires a different way of thinking and responding to the world."

In essence, the *speed of making good decisions* is the critical element in our strategies. Speed, however, is something that is not always fully understood. Speed can be obtained by loosening up the command-and-control structure, allowing for more chaos in the decision-making process, and unleashing the creativity of front-line workers. Speed is not necessarily a function of technology. In war, technology can actually slow an army down. Look at the North Vietnamese versus the U.S. in the Vietnam War—The poorly-armed North Vietnamese fighters took on America's technological military might with lightning-fast guerilla attacks.

Those who will succeed in the current hyper-competitive global environment know how to embrace change. They *unlearn* their former operating models to stay rooted in the OODA Loops. It's their mental models that create advantage, not the size of their assets, number of employees, or number of trading partners.

If a company currently enjoys success, watch out, for there's little time to enjoy success in today's ruthless times. The winning company will re-Observe, re-Orient, re-Decide and re-Act at all times. Call this approach "sense and respond" if you wish to keep it brief.

The convergence of total global competition, global hyper-connectivity, and dynamic outsourcing and supply chaining has changed the meaning and practice of strategy. Walt Shill, former McKinsey consultant and now with Accenture famously stated, "*Strategy, as we knew it, is dead. It's now all about operational agility and how fast businesses can seize opportunities. If strategies and forecasts have to change daily or weekly, then so be it.*" We've moved from *return-on-assets* to *return-on-opportunities,* and those opportunities are in constant flux and under constant threat in a constantly changing environment.

R.I.P
STRATEGY
d. 2010

We can turn to nature to learn lessons about competing on *time* as the central variable in the OODA Loop. As the authors of *The Real-Time Enterprise* write, "Polar bears have a unique ability to smell (sense) through thin ice to locate their favorite prey, the seal. But they also know that their prey is an elusive target, and that it's knowing exactly when to pounce (respond) through the ice that counts. Pouncing too soon, before maneuvering to the right spot where the ice will break cleanly, or pouncing too late after competitive bears have spotted the prize, will surely mean lost opportunity and continued hunger. It's all in the timing—it's time-based competition for survival in a harsh world." Timing and agility go hand in hand.

tinyurl.com/62g2qjn

McKinsey & Company maintains that competitive advantage consists of the progress a company makes as its competitors, paralyzed by confusion, complexity and uncertainty, sit on the sidelines.[2] The key is to be ready to pounce on an opportunity as soon as the company can realign the business processes of its value delivery system, not just to be the first mover, but also to be able to swim hard once it has broken the ice with innovation. Having caught up with the bear and joined in the melee, competitors will only have remnants of the prize to fight over while the innovator moves on, scouting for new opportunities to surprise both its prey and its competitors.

Barriers that slow down responsiveness and adaptation include non-value-added processes, disconnected departmental hand-offs, or anemic business processes. An anemic bear simply cannot compete in the harsh real world—time-based competition demands operational transformation through business process innovation. Timing and agility go hand in glove.

Just consider Nokia, Apple, Samsung and Motorola and the other competitors slogging it out in the ruthless smart phone and tablet arenas. It seems that the innovator, Apple, dominates the tablet market while the others squabble over the remaining scraps. Hammonds also wrote in *Fast Company*, "Companies in manufacturing, telecommunications, retail—in nearly every business—are discovering that fashion, fad, and fickle customers require constant vigilance and adjustment. We operate in a video-game world where time is compressing, information goes everywhere, and the rules of the game change abruptly and continuously. All of which makes the OODA loop more powerful than ever. Want to out-think and out-execute the competition in the air or on the ground, in combat or in business? Want to test out new ideas, get feedback from your customers, adjust your product accordingly, and launch a new version—before your competition even senses the opportunity? Then learn how to make the OODA loop the centerpiece of your strategy process." [3]

"In Boyd's notion of conflict, the target is always your opponents mind," says Grant Hammond, director of the Center for Strategy and Technology at the Air War College. In his own work, Boyd didn't apply his principles to business strategy and market share, says Hammond, "But the analogy still holds. It's all about rapid assessment and adaptation to a complex and rapidly changing environment that you can't control." In fact, Boyd's ideas translate seamlessly into business. In a groundbreaking article published in 1988 in the *Harvard Business Review* titled "Fast-Cycle Capability for Competitive Power," Joseph L. Bower of Harvard Business School and Thomas M. Hout, a partner at Boston Consulting Group, cited the OODA loop, "The OODA loop limbers up your organization. It keeps you constantly worried about the next cycle, about making rapid, incremental improvements that throw off competitors."

But in business, it's not one big cycle that needs to be managed, it's multiple overlapping cycles as shown below in a figure from the book, *The Real-Time Enterprise.*

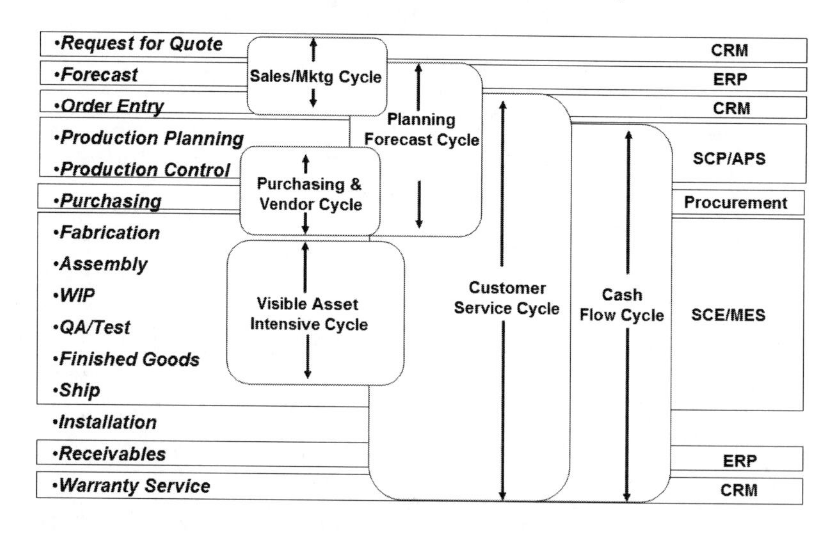

The book's chapter on cycle-time management opens with, "Cycle-time management is essential to the effectiveness, agility and overall productivity of any company. Reducing information float, increasing information synchronization and achieving near zero information latency are the ingredients of cycle-time management. Over the years, IT has provided automation support for discrete business functions. Although these business functions (listed on the left of the figure above) are vital, they are not how a company runs its business.

What companies want to do is manage their asset-intensive business cycles. They want to manage their purchasing cycle. They want to manage their forecast and planning cycle, the customer service cycle and other business cycles, as illustrated in the figure.

Each major business cycle is a complete, end-to-end business process, usually requiring support from more than one company to realize the process." Each major business cycle requires its own OODA Loops as well as the overall orientation that influences all the loops at strategic and tactical levels.

Hammonds continues, "Toyota, studied by Boyd and others, designed its organization to speed information, decisions, and materials through four interrelated cycles: product development, ordering, plant scheduling, and production. Self-organized, multifunctional teams at Toyota developed products and manufacturing processes in response to demand, turning out new models in just three years compared with Detroit's cycle of four or five.

"Systems like Toyota's worked so well, Boyd argued, because of schwerpunkt, a German term meaning organizational focus. Schwerpunkt, Boyd wrote, 'represents a unifying medium that provides a directed way to tie initiative of many subordinate actions with superior intent as a basis to diminish friction and compress time.'

"That is, employees decide and act locally, but they are guided by a keen understanding of the bigger picture. In effective organizations, schwerpunkt connects vibrant OODA loops that are operating concurrently at several levels. Workers close to the action stick to tactical loops, and their supervisors travel in operational loops, while leaders navigate much broader strategic and political loops. The loops inform each other. If everything is clicking, feedback from the tactical loops will guide decisions at higher loops and vice versa." (See also, "Fractal Enterprise Architecture:" http://tinyurl.com/4kgax6s).

"Think of the loop as an interactive web with orientation at the core. Orientation—how you interpret a situation, based on your experience, culture, and heritage—directly guides decisions, but it also shapes observation and action. At the same time, orientation is shaped by new feedback.

"An effective combatant, Boyd reasoned, looks constantly for mismatches between his original understanding and a changed reality. In those mismatches lie opportunities to seize advantage. And reality, Boyd understood, changes ceaselessly, unfolding 'in an irregular, disorderly, unpredictable manner,' despite our vain attempts to ensure the contrary. 'There is no way out,' Boyd wrote. 'We must continue the whirl of reorientation, mismatches, analyses/synthesis over and over again ad infinitum.' The OODA loop persists endlessly."

"On the other hand, it may be that technology compresses just one part of the loop, that the wide, instantaneous availability of data creates an environment of complete transparency. In such a world, it would be impossible to gain advantage from observation as all competitors would see the same thing. *Orientation*, then, becomes even more important: The data is worthless, after all, without our interpretation. And that means Boyd was more right than even he could have imagined."

Robert Lewis, president of IT Catalysts, brings some balance to the discussion of OODA Loops, "As generally conceived, an OODA Loop is used to win the battles you fight, but not necessarily the way to choose the wars you wage. And in fact, one way of understanding what went horribly wrong in the second Iraq war is that the folks who decided it was a good idea were focused on their OODA-driven ability to win it (and, one might speculate, their OODA-driven ability to sell it), leading them to ignore the question of whether it made sense to fight the war in the first place. Had they allowed some of that analysis paralysis to take place they'd have figured out that removing Saddam Hussein from power would eliminate the main counterbalance to Iran in the region."

Because we are dealing with multiple OODA Loops in any complex system such as a business, we need to keep front and center the notion of schwerpunkt to provide overall focus. Strategic OODA Loops may indeed cycle slower than tactical loops, and a lot more emphasis may be needed in the Observe phase.

Strategic as well as tactical OODA Loops require continuous cycling regardless of their speed; feedback loops are all in a world of continuous change. If a leading manufacturer of 8-track tapes didn't Observe the megatrend that was about to hit it (cassette tapes), it may have continued to tactically compete against other 8-track tape companies instead of getting the hell out of Dodge City. Ditto for the compact disk makers with the advent of MP3 players.

Florida attorney, Stratton Smith, would like to teach a MBA course with a one-question syllabus, "As CEO of a major energy company, what are you going to do when someone invents a radically low-cost, clean, scalable, distributed source of energy? Get back to me with your scenarios at the end of the semester. Your grade and your future in business depend on your results."

Key to Smith's MBA challenge is the word "when," for with 7 billion brains on the planet, the question is certainly about *when*, not *if*. Smith, of course, is asking the class to develop a bundle of possible scenarios that can be drawn from to unleash the appropriate quick-response OODA Loops when disruptions happen rather than being shocked and awed when shift happens. Call it pre-planning for possible futures, or use the management techniques of scenario-based planning or pattern-based strategy.

Pattern-based strategy doesn't mean putting a group of geniuses in a room for a week. It means gaining the analytic capabilities needed to tap social networks with "listening posts" to make sense of what's now tagged as "Big Data," the zillions of bits of unstructured information pouring out of blogs, emails, videos, global social networks. Theses ideas are explored further in Chapter 6.

For now, let's just recap that tactical OODA Loops require schwerpunkt, or organizational focus, that in turn must be informed by going to the *edge* of our rapidly changing world to maintain a bundle of possible scenarios in our competitive strategy portfolio.

Returning to the MBA class, the students would need to go to the edge to pick up on information such as this from MIT, "Major discovery from MIT primed to unleash solar revolution. Scientists mimic essence of plants' energy storage system."[4]

Daniel G. Nocera, Professor of Energy at MIT and his team have developed a simple method to split water molecules and produce oxygen gas, a discovery that paves the way for the large-scale distributed use of solar power. What's a coal-burning energy executive to do when artificial photosynthesis scales and goes global?

In a revolutionary leap that could transform solar power from a marginal, boutique alternative into a mainstream energy source, Nocera and his team have overcome a major barrier to large-scale solar power: storing energy for use when the sun doesn't shine. Until now, solar power has been a daytime-only energy source, because storing extra solar energy for later use is prohibitively expensive and grossly inefficient.

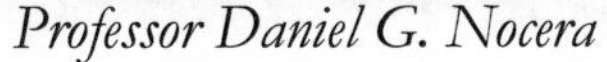

Professor Daniel G. Nocera   *Watch: tinyurl.com/6ruyrc5*

The MIT researchers have hit upon a simple, inexpensive, highly efficient process for storing solar energy. Requiring nothing but abundant, non-toxic natural materials, this discovery could unlock the most potent, carbon-free energy source of all: the sun. "This is the nirvana of what we've been talking about for years," said Nocera, the Henry Dreyfus Professor of Energy at MIT and senior author of a paper describing the work in the July 31, 2008 issue of *Science*. "Solar power has always been a limited, far-off solution. Now we can seriously think about solar power as unlimited and soon."

Inspired by the photosynthesis performed by plants (biomimicry), Nocera and Matthew Kanan, a postdoctoral fellow in Nocera's lab, have developed an unprecedented process that will allow the sun's energy to be used to split water into hydrogen and oxygen gases. Later, the oxygen and hydrogen may be recombined inside a fuel cell, creating carbon-free electricity to power your house or your electric car, day or night.

The key component in Nocera and Kanan's new process is a new catalyst that produces oxygen gas from water; another catalyst produces valuable hydrogen gas. The new catalyst consists of cobalt metal, phosphate and an electrode, placed in water. When electricity—whether from a photovoltaic cell, a wind turbine or any other source—runs through the electrode, the cobalt and phosphate form a thin film on the electrode, and oxygen gas is produced. Combined with another catalyst, such as platinum, that can produce hydrogen gas from water, the system can duplicate the water splitting reaction that occurs during photosynthesis. The new catalyst works at room temperature, in neutral pH water, and it's easy to set up. "That's why I know this is going to work. It's so easy to implement," Nocera said.

Sunlight has the greatest potential of any power source to solve the world's energy problems. In one hour, enough sunlight strikes the Earth to provide the entire planet's energy needs for one year.

"This is a major discovery with enormous implications for the future prosperity of humankind," said James Barber, the Ernst Chain Professor of Biochemistry at Imperial College London. "The importance of their discovery cannot be overstated since it opens up the door for developing new technologies for energy production thus reducing our dependence for fossil fuels and addressing the global climate change problem."

It wasn't so long ago that we became a "Hydrocarbon Society" and, in the language of anthropologists, we became "Hydrocarbon Man."

By the way, keep an eye out for the next-generation battery. If you haven't heard of Professor Ann Marie Sastry by now, you've probably been spending too much time and money at the pump putting gasoline into your car.

Sastry is the CEO and co-founder of a hot, new automotive battery development company. In 2007, she founded a startup called Sakti3 to develop solid-state (versus liquid) batteries that don't require most of this added bulk. They save even more space by using materials that store more energy. The result could be battery systems half to a third the size of conventional ones. Cutting the size of a battery system in half could cut its cost by as much as half, too.

Both professors Nocera and Sastry realize that their "inventions" are worthless without "business innovation," and have turned to the likes of GE, Tata and Kholas Ventures among others to transform their *inventions* into *business innovation.* Both work within Boyd's OODA Loops. "To be very honest, it depends on a lot of things," Sastry says. "Depends on how fast we run [OODA Loops], depends on the dollars that come in, it depends on how successful and, sometimes, how lucky you are in doing the technology. So, there are a lot of variables. You know, starting a new business is very risky."

Watch Sastry: tinyurl.com/6pouqo5

In a profile on Sastry, journalist Matthew Dakotah wrote, "Sastry approaches the realities of entrepreneurship with realism, but she clearly sees a path to success for her new company. 'We may fail. That means that we're taking appropriate risks. And as far as the competitors are concerned, I certainly hope they're working as hard as we are,' she says. 'I don't mean that as a throw down. We've got huge numbers of people in the emerging economies that are going to join the middle class and they may adopt the internal combustion engine [instead of electric vehicles] unless the science and technology fields are working hard on energy storage. The markets are enormous and there is room for dozens and dozens of companies to fill the need.'"

"And how will all of those people join the middle class? By having parents that set the same kind of expectations that Sastry's father did. 'When you look at the numbers of people going into technology fields globally, they dwarf our own numbers. In prior decades the United States had hegemony in math, science and technology," she says. 'It's fading because other nations are becoming very savvy to the fact that people who offer unique capabilities in science and technology are in high demand, and can command higher salaries and create a better way of life for their families.'"[5]

Nocera and Sastry may not have to rely on solar cells or wind turbine farms to power the artificial leaf or top up the next-generation battery. An interesting "Windstalk" concept devised by New York design firm, Atelier DNA, could overcome the problems of wind turbine farms while still allowing a comparable amount of electricity to be generated by the wind. Instead of relying on the wind to turn a turbine to generate electricity, when the poles (stalks) sway in the wind, a stack of piezoelectric discs are compressed, generating a current through the electrodes. In a nice visual way to indicate how much, if any, power the poles are generating, the top 50cm (20 in.) of each pole is fitted with an LED lamp that glows and dims relative to the amount of power. So when the wind stops, the LED's go dark.

To maximize the amount of electricity the Windstalk farm can generate, the concept also places a torque generator within the concrete base of each pole. As the poles sway, fluid is forced through the cylinders of an array of current generating shock absorbers to convert the kinetic energy of the swaying poles into electrical energy.

Because the electricity generation capabilities of a Windstalk field site would depend on the wind, the designers have devised a way to store the energy. Below the field of poles are two large chambers located on top of each other and shaped like the bases of the poles but inverted. When the wind is blowing, part of the electricity generated is used to power a set of pumps that moves water from the lower chamber to the upper one. Then, when the wind dies down, the water flows from the upper chamber down to the lower chamber, turning the pumps into generators. Shown below is a pilot Windstalk farm devised as a potential clean energy generation project and tourist attraction for Abu Dhabi's Masdar City.

Read more: tinyurl.com/3aj5dz5

Caveat oil companies. You'd better turn up the heat on your business agility burners to sense and respond to these disruptive changes in the world of energy.

The key to gaining true business agility isn't just about mastering techniques related to OODA loops, it's *cultural.* Ajit Kapoor, IT industry veteran and former Enterprise Architect at Lockheed wrote, "This reminds me of a meeting that Gandhi and God may have had some time back, say 50 years after his demise from earth. Frustrated with the situations in India, Gandhi asked God, 'Why are Indians, who loved him dearly, then are not listening to what he gave his life for?' God replied and asked Gandhi if he would like to go back and fix the problem. But he added further that it's like a movie set, "Scene 1, take 1 million. See Mr. Gandhi, humans do not change easily. They continue with their old ways and expect different results."'

In his book, *Strangers to Ourselves: Discovering the Adaptive Unconscious,* Timothy D. Wilson, Professor of Psychology at the University of Virginia, explains that it's not so easy to let go of our old ways, "The unconscious mind plays an intriguing role in processing a stimulus in the present moment and connecting it to a memory that is then brought up for recall. This process is done instantaneously and without our control.

"Because we don't have access to our unconscious processes, the conscious mind has become good at creating reasons for why we act or think the way we do. We rely on four sources to help us create these reasons: shared cultural theories; prior experiences and explanations; idiosyncratic theories; and private thoughts, feelings and memories. Without being aware of the process, the unconscious mind is continually drawing inferences about what we like, who we are, and what we want to become, and then taking that information to shape how we act in the present. Though we cannot shut this process off, we can gain better control of it if we set our conscious mind to do the same thing, thus synchronizing the two minds."

In business or government we can turn to the words of professor and author, Clay Shirky, "Institutions will try to preserve the problem to which they are the solution." In a way, the Shirky Principle is similar to the Peter Principle, which says that a person in an organization will be promoted to the level of their incompetence. At this point their past achievements will prevent them from being fired, but their incompetence at this new level will prevent them from being promoted again, so they stagnate in their incompetence. The Shirky Principle declares that complex solutions (like a company, or an industry) can become so dedicated to the problem they are the solution to, that often they inadvertently perpetuate the problem.

OODA Loops in the Cloud

What's all this OODA Loop activity got to do with business innovation and the Cloud? Everything! Let's quickly explore the key Cloud delivery models and their effects on *executing* on business innovation:

■ *Infrastructure as a Service (IaaS)* - Even startups requiring the power only supercomputers can provide are able to deploy the resources of massive data centers without one dime in capital investment. With funding from family and friends, Animoto was started by some young techies that worked for MTV, Comedy Central and ABC Entertainment who knew how to make professional quality video animations. Now their Cinematic Artificial Intelligence technology that thinks like an actual director and editor and high-end motion design bring those capabilities to anyone wanting to turn their photos or videos into MTV-like videos. At one point, aside from some monitors and an espresso coffee machine Animoto had few actual assets. That's because everything, including server processing, bandwidth and storage, is handled by cloud computing, a pay-as-you-use model. So when the Animoto application launched on Facebook, causing the number of users to soar from 25,000 to 750,000 in four days and requiring the simultaneous use of 5,000 servers, business carried on as usual. Without the ability to handle a spike like that, their business couldn't exist.

Meanwhile, it's not just youngsters using IaaS. The *New York Times* (certainly no startup) processed four terabytes of data through a public cloud by simply using a credit card to get the service going. In a matter of minutes it converted scans of more than 15 million news stories into PDFs for online distribution—at an incredibly low cost of$240! Look Ma, no New York Times IT infrastructure needed. Meanwhile, Nasdaq uses public cloud storage to deliver historical stock and mutual fund information, rather than add the load to its own computing infrastructure.

■ *Platform as a Service (PaaS)* - With PaaS, software developers can build Web software without installing servers or software on their computers, and then deploy the software without any specialized systems administration skills. PaaS service providers not only incorporate traditional programming languages but also include tools for mashup-based development, meaning that deep IT skills are not needed to build significant software. The implications for business innovation center on rapid development and rapid testing, making it possible to bring new products and services to market without the traditional 18-month IT development cycle or capital expenditures. Innovations that don't pan out can be shut down, allowing a company to fail early, fail fast. Remember, innovation must allow for failure, or else nothing really *new* is being done. On the other hand, innovations that prove successful can be scaled up to full Web scale in an instant. In short, PaaS takes traditional IT software development off of the critical path of business innovation.

■ *Software as a Service (SaaS)* - With SaaS we are witnessing a huge shift from IT to BT (Business Technology). In the past, IT was about productivity. But now, BT is about collaboration, a shared information base and collective intelligence (the wisdom of crowds, social networking and crowdsourcing). SaaS is the delivery of actual end-user functionality, either as "services" grouped together and orchestrated to perform the required functionality or as a conventional monolithic application (e.g., CRM, ERP or SCM). The real driver for SaaS isn't the traditional IT application; it's the "edge of the enterprise" where business users require a flexible model to deploy new technologies to improve front-office performance. On the other hand, companies are't going to discard their existing enterprise systems, they, instead are going to leverage them and integrate them into the Cloud services they wish to provide.

As a growing number of business units tap SaaS offerings without going through their central IT department, we have the advent of "Shadow IT." The key significance is that while IT has a major role in the enterprise back office (transaction processing and systems of record), these new requirements are directly associated with "go-to-market" activities and will be subject to constant change via OODA loops. These new requirements must be met very quickly for competitive purposes; some are likely to endure for only a few months; and their costs will be directly attributed to the business units consuming the needed "services" and paying as they go.

Now consider operational innovation inside a huge company like GE blending both internal clouds and going beyond the firewall to reach out to suppliers in the Cloud. GE's supply chain is huge, including 500,000 suppliers in more than 100 countries that cut across cultures and languages, buying up $55 billion a year. GE wanted to modernize its cumbersome home-grown sourcing system, the Global Supplier Library, build a single multi-language repository, and offer self-service capabilities so that suppliers could maintain their own data.

So did CIO Gary Reiner and team start programming? The short answer is "no." GE looked to the Cloud for a solution. GE engaged SaaS vendor Aravo to implement its Supplier Information Management (SIM) that would ultimately become the largest SaaS deployment to date. GE is deploying Aravo's SaaS for 100,000 users and 500,000 suppliers in six languages. When GE goes outside its firewall to innovate, you can bet that other CEOs will be asking their CIOs lots of questions about harnessing the Cloud for operational innovation.

■ BPM as a Service (BPMaaS) - Business Process Management (BPM) is what sets "enterprise cloud computing" apart from "consumer cloud computing." Because the average end-to-end business process involves over 20 companies in any given value chain, multi-company BPM is essential to business innovation and maintaining competitive advantage. Bringing BPM capabilities to the Cloud enables multiple companies to share a common BPM environment and fully participate in an overall end-to-end business process. BPMaaS can be implemented as a "horizontal" Business Operations Platform (BOP) that has a Business Process Management System (BPMS) at its heart. This is similar to PaaS, but rather than programming tools being accessed, the BPMS is being accessed for full process lifecycle management and specific process services such as process modeling and business activity monitoring.

For example, using a Business Operations Platform from Cordys, Lockheed Martin has deployed a Cloud-based Collaborative Engineering system to orchestrate the work of hundreds of subcontractors that have disparate product lifecycle management and CAD /CAM systems. This represents one of the world's most complex enterprise computing environments now being addressed by cloud computing. Meanwhile, Dell, Motorola, Boeing, Avon, Panasonic, IBM and other multinationals use e2Open's Business Network to provide complete demand and supply chain management in the Cloud.

Nowhere is the OODA Loop more applicable than in supply chain management, especially if you consider the massive disruptions that resulted from the tsunami in Japan or the need to bring new products and services to market with great speed. While BPMaaS can enable companies to manage business processes more efficiently, its real business innovation impact is that it can also empower entirely new business models that dynamically integrate demand-supply chain partners into virtual enterprise networks that offer compelling value.

Jasmine Young, a Facilitator at the Haas School of Business Institute for Business Innovation, summarized, "The Cloud is about leverage, the way credit is leveraged in the financial industry. Businesses need to think about how they can leverage their suppliers and partners—and customers. And that's how the case toward innovation in the Cloud can best be driven." By aggregating more and more offerings for their customers, industry boundaries become blurred as smart competitors enter markets outside their primary industries. ExxonMobil is in the gourmet coffee business. Starbucks is in the Internet business. Walmart is in retail banking. Microsoft is in the telephone business with its acquisition of Skype.

We could spend hours exploring how the OODA Loop fits into cloud computing or embracing the Cloud for business innovation. Or we could compare it to the Plan, Do, Check and Act (PDCA) model originating with quality management guru, Edwards Deming. According to the author and veteran enterprise architect, Thomas Tinsley, "For Deming the outcome was improved quality, where OODA is about survival."

All this OODA Loop activity happens in the Cloud *leveraging* IaaS, PaaS, and SaaS capabilities, for it's not Industrial Age assets that must be managed, it's *digital immediacy* and the weaving of a digital tapestry among our customers and trading partners that counts in 21st Century dogfights of business innovation.

Takeaway

Leading companies are taking the unsystematic approach to business innovation and turning it into a repeatable, managed business processes—think Innovation Process Management (IPM). IPM can be compared to the rise of the total quality movement in the1980s, where leaders such as Toyota taught the lesson of quality-or-else. Some companies have already implemented systematic approaches to innovation management. GE calls it CENCOR (calibrate, explore, create, organize and realize) [what's the "N" for?] and it centers on Design for Six Sigma. The Mayo Clinic calls it

SPARC (see, plan, act, refine, communicate). The design firm, Doblin, uses an Innovation Landscape™ diagnostic method to show 10 types of innovation and reveal that the most sophisticated innovation strategies combine these in thoughtful ways.

Although leading companies and innovation consultants have many innovation process roadmaps, OODA Loops and Energy-Maneuverability theory provide the baseline, a unifying theory of *agility*, for any business innovation process worth its salt.

References.

---

[1] http://www.fastcompany.com/magazine/59/pilot.html

[2] McKinsey Quarterly, 2002 Number 2, *Just-in-Time Strategy for a Turbulent World*

[3] http://www.fastcompany.com/magazine/59/pilot.html

[4] http://web.mit.edu/newsoffice/2008/oxygen-0731.html

[5] http://www.huffingtonpost.com/matthew-dakotah/ann-marie-sastry-bio-profile_b_884639.html

6. Innovating Innovation Itself

Larry Keeley, Doblin

"Since innovation fails about 96% of the time, it seems self-evident that the field has advanced to about the same state as medicine when leeches, liniments and mystery potions were the sophisticated treatments of the day." —Larry Keeley, Doblin. On the occasions when Larry can get someone to listen, he is inclined to reveal pieces of the emerging "science of innovation."

According to *Business Week's* Bruce Nussbaum, "There is, in fact, a whole new generation of innovation gurus. They are not the superstars of the '90s, such as Clayton Christensen, who focused on what might be called macro-innovation—the impact of big, unexpected new technologies on companies. The new gurus focus more on micro-innovation—teaching companies how to connect with their customers' emotions, linking research and development labs to consumer needs, recalibrating employee incentives to emphasize creativity, constructing maps showing opportunities for innovation."

But there's more to innovating innovation; it's about innovation as a *systematic and repeatable business process.* The man who just might turn out to be the Edwards Deming of 21st century innovation, Howard Smith, CTO of Computer Sciences Corporation, thinks business innovation is really about *process.* Where Deming brought process to the quality movement, Smith goes beyond the notions of creativity, invention and design to bring a rigorous problem-solving process to innovation. "Reflecting on the invention of the Alto personal office computer, author, consultant and former Director at Xerox PARC labs, John Seely Brown observes that 'as much, if not more, creativity goes into the implementation part of the innovation as into the invention itself.' In this respect, Xerox, the inventor, failed as an innovator, leaving billions in profits for Apple and Microsoft. Creativity, invention, design and business innovation are often confused."

Smith explains, "Innovation is a holistic process involving the entire organization of a commercial enterprise, whereas invention is a discrete event, typically performed by specialist individuals or very small teams. Innovation requires multi-disciplinary teams and is a complete lifecycle process. Creativity and design are necessary, but insufficient. In this sense, IDEO's [one of the world's great design firms] design innovations are, like every other element in the *operating system for innovation,* a part of the mix. Yet in a world of product abundance, mass-customization and extraordinary high expectations when consumers interact with public or private services or business people deal with suppliers, IDEO's core competence is no doubt a vital ingredient. Their design process turns genuine inventions into useable, interesting and beautiful products and services, rendering them acceptable to commercialization. And what IDEO produces must be relevant to markets, and the timing of the release of those innovations to markets is critical, as Christensen has taught us. Yet just as we must move beyond Christensen's management frameworks if we are to understand the sources of innovation and the critical role of problem solving, so too must we move beyond IDEO's design innovation if we are to understand the full extent of what innovation is. Seen as the creator of new value, innovation isn't hit-or-miss, trial-and-error lateral thinking, but a repeatable process. What is innovative about innovation today is the realization that it can be achieved systematically, and that the innovator is an obsessive *problem solver.*"

To put teeth into his approach of business innovation as problem solving, Smith goes way beyond the many techniques and methods most often associated with innovation, "But if you thought you had heard about all the best-practice acronyms and trends out there, think again. To the current plethora of strategies for adaptation and survival is now added something that may be a way of thinking, a set of tools, a methodology, a process, a theory or even possibly a deep science, but which may be gradually shaping up as 'the next big thing.' It's called TRIZ, pronounced 'trees' and is an acronym for the Russian words that translate as 'The Theory of Inventive Problem Solving.' Its systematic approach to innovation is the antithesis of unreliable, hit and miss, trial and error, psychological means of lateral thinking. Its scientific, repeatable, procedural and algorithmic processes surprise all who first encounter them. Sound like magic?

"After just one TRIZ workshop, engineers at National Semiconductor modified a machine that tests integrated circuits (ICs) that had gobbled up $76,000 in the previous five months of trial and error. Within a week, TRIZ-based software responded with 40 directions in which the engineers could investigate a solution. The most promising idea was the replacement of frail IC contacts with an elastomer, reducing the physical impact to IC leads during insertion. The consensus among the engineers working on the problem at the time was that, without guidance from TRIZ, the project would still have been hunting for a solution. Other companies have had similar experiences.

"As globalization advances and companies see fewer opportunities for growth, the clamor for invention and innovation—proxies for 'economic value'—will inexorably rise. Innovation poster-child GE redid a twenty three year old slogan called, 'We bring good things to life,' and replaced it with a slogan called, 'Imagination at work.' The firm includes a creative drawing tool on its Web home page. By contrast, FedEx is almost dull. Its core competence in logistics implies supply-chain efficiency and reliability. Those qualities define the FedEx 'identity' business process.

"Is FedEx less innovative than GE? Not necessarily. What do GE and FedEx have in common? Both are obsessive problem solvers.

"Companies do more than perfect the known and optimize for efficiency. Glib use of the terms 'creativity' or 'innovation' means little if relevant problems are not being solved. Innovative firms develop an ability to solve problems that remove barriers to greater economic value. Whether an engineer is figuring out why an industrial process won't start, or a call center operator is re-designing support processes to avoid answering similar problems over and over again, both are solving problems and each requires methodology and in-context expertise. At the macro level, numerous elements are involved: a learning environment, creative thinking tools, design flair, engineering skill, scientific method, enabling work practices, an amenable culture, specific organizational structures, supportive management frameworks, numerous business processes, information systems, market strategy, and predictive algorithms. At the micro level it comes down to the individual employees, their talent, qualifications and knowledge."

An Architecture for Innovation

"Sustainable innovation can be achieved through a managed process aimed at resolving persistent business issues, creating new business models or designing products and services that address unfulfilled customer needs. A well-managed innovation process can be designed to elicit ideas that are highly relevant to an organization's specific or broad business objectives. In short, innovation can become a managed, reliable and fruitful process."[1] —Gartner Five Myths of Innovation, 2010 Update, Carol Rozwell and Kathy Harris

How do you make innovation a systemic and repeatable process? The starting point is having an architecture or operating environment from within which innovation can take place. This is the same as having a business or enterprise architecture for running your company.

An architecture is the use of abstractions and models to simplify and communicate complex structures and processes to improve understanding and forecasting the behavior of the system – in this case the system of innovation. Architecture uses sets of abstractions and models of an environment, problem space or domain, either physical or logical, with a set of associated views into that domain to provide:

- Simplification and management of complexity in all of its forms (structural, procedural or informational), in particular the management, understanding and integration of the business and technical domains.
- Communication and common understanding of the problem space to multiple stakeholders from widely different environments by the use of multiple domain specific views of the architectural model.

- Completeness and relationship analysis of proposed solutions in the problem space or domain by examining the models and architectures from multiple differing viewpoints for incompleteness and gaps.
- Forecasting and predicting future architectures, strategies, structures, patterns, relationships and technologies in the business and technical space by extrapolation of abstractions and models.

In other words in relation to buildings, architecture has to do with the planning, designing and constructing of form, space and ambience that reflect functional, technical, social, environmental, and aesthetic considerations. Likewise from an organizational perspective, an enterprise architecture (EA) is a rigorous description of the structure of an enterprise, which comprises enterprise components (business entities), the externally visible properties of those components, and the relationships (e.g. the behavior) between them. This description is comprehensive, including enterprise goals, business processes, roles, organizational structures, organizational behaviors, business information, software applications and computer systems, as well as terminology used and guiding principles for changing itself (self-similar, fractal and self –referential).

While, "At the micro level it comes down to the individual employees, their talent, qualifications and knowledge," that is the case for all work. What you want to create is an environment where the individual's work is eased for efficiency and effectiveness (ideas aren't being implemented if the staff is running around trying to figure out who to tell) and for leverage and diffusion through the organization.

Here is a very high-level and simplified example of an Innovation Architecture:

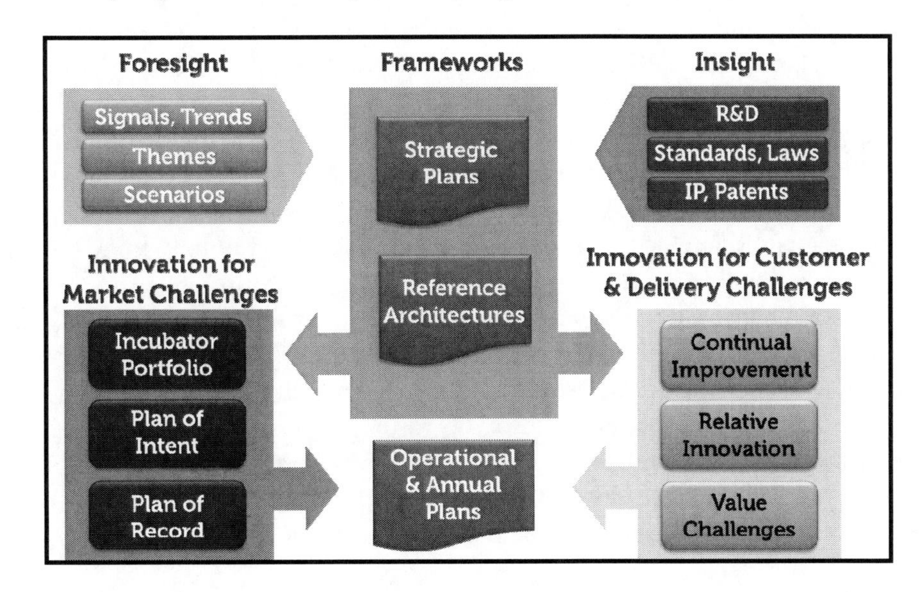

It is not meant to illustrate all of the connections, inter-dependencies, components, and interactions, nor is it a definitive process slide for an innovation management system. It is illustrative as a starting point for developing your own.

The key is to maintain a multi-dimensional and divergent (creative, intuitive, holistic, conceptual) view (foresight) of all the trends and ideas in as many domains as possible, translate those into the value systems and requirements (themes) that will be used for making decisions in the future and then construct stories based upon those requirements and values. Likewise, you have to maintain a broad view (insight) and stay rooted within the constraints in place (because changing status quo is hard). You'll also have to break away from the "practical, convergent, disciplined" (analytical, quantitative, sequential, specific) thoughts of your industry and markets.

And all of this has to be available to your teams 7x24x52 so they know where you need fresh ideas, how to communicate their fresh ideas, how to test those great ideas, and then how to execute on them.

Foresight and Insight

Somewhere in the organization someone has to be monitoring the outside world, what is it doing, where is it going, who is out there… all the elements of scanning identified in "Observe" in OODA. Identify trends, using for example the STEEP Framework (Society, Technology, Environment, Economy, and Politics), that cause change resulting in challenges and opportunities for you and your customers.

If you want to be innovative, you want to understand that the ability to innovate is driven by your OBSERVEation and your ORIENTation capabilities. The key is your ability to understand social, economic, business and regulatory trends, and for formulating scenarios of where they're going to end up, and your ability to apply technology to the problems and opportunities uncovered.

Foresight: A foresight model anticipates and monitors changes in markets, economies, regulations, demographics, business models, processes, values, priorities, etc. – all of which would benefit from enablement, facilitation and acceleration from technology. Technology trends are also investigated to ensure what is possible is understood in order to establish probable time lines for the anticipated changes. The changes to be looked at are not just products and services but new businesses, new business models, new business processes—and the sea change of technology transformations that are going on outside of your specific industry that will radically affect your industry.

Foresight should be focused as far into the future as possible, it should identify thought leaders (both in and out of your domain), and it should be looking for patterns and the emergence of trends very early on, sometimes called weak indicators. It requires heavy immersion into the information stream noticing what is being said, and what is no longer being said. The key to foresight is not to predict an innovation or new technology or new need, but to describe what future buyers will value, what their decision processes and criteria will be, what expectations will exist (how many companies introduce a new product only to find the market has moved on?), what conditions and environments for buying and consuming might be like.

That foresight is then translated into a series of themes that describe the high-level or meta-level behavior of the future. Those themes are used to construct scenarios that can be used for strategic product and services planning. These scenarios are really stories about the future; they paint a picture of what might be and how the world would operate if that came to be – like a good science fiction short story. Additionally scenarios provide a framework wherein the inter-relationships of the scenarios can be used for investigating and positioning the potential for innovation. Consider scenarios related to the Internet of Things that drives Big Data; Big Data, in turn, drives embedded real-time actionable business intelligence which in turn drives the need for multi-sensory man-machine interfaces like the AlloSphere.

Insight: In some ways insight is the opposite of foresight. It is based upon the linear trends; the best practices; what customer surveys say; what the industry gurus and analysts say. It also looks at what would constrain the course of action or degrees of freedom that are available for innovation. This could include both de jure and de facto standards and practices, current business assumptions, and the current legal and regulatory environment. Insight also looks at the not-so-distant future that is more easily predictable in terms of where the potential breakthroughs are. What is coming out of your own and others labs and research? What Proof of Concepts and trials are going on in the market? In many ways, insight is what most companies have done in preparation for their strategic planning in a convergent thinking manner.

A Caution. One can get carried away with this process thing. It is there to support and enhance the "creativity" thing and that sometimes gets lost. The purpose of the FORESIGHT process is to create and encourage creative, divergent thinking. The data, information, outcomes and analysis should be creative, intuitive, qualitative, subjective, reflect possibilities (what could or might be). They should represent a holistic view (not just a narrow perspective—how many buggy whip manufacturers were thinking about the transportation industry and were prepared for the automobile?), and generally should generate conceptual abstractions. On the other hand, foresight needs to be balanced by the INSIGHT—very analytical, rational, quantitative and sequential analysis and convergent thinking around what can be seen *tangibly*. The information from insight should be driven by *constraints* (what cannot be done, at least from a rational perspective), should be objective and should always be backed up with specific details.

It is then combining those two different views of foresight and insight that you now can be truly observant (OBSERVE in OODA) of all that is going on and can orient (ORIENT in OODA) yourself to the best course of action and the right form of innovation.

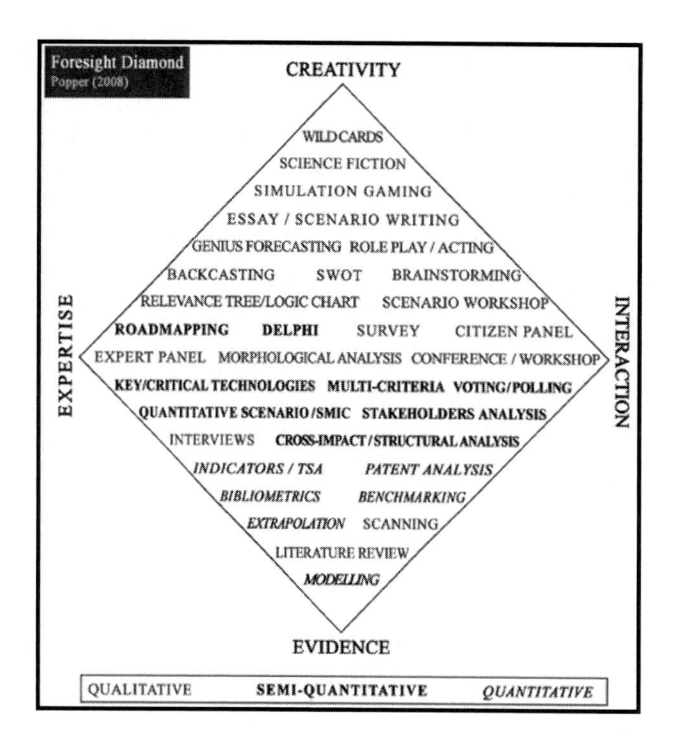

Frameworks: The New Role of Strategy

When thinking about the future, how do we avoid what the economists and behaviorists call the "horseless carriage syndrome;"[2] referring to how early automobiles tried to be as familiar to horse-drawn carriages as possible, including the addition of whip holders and even fake horse heads? At the same time can we avoid "blinkered views,"[3] an inability to get out of one's own mind-set and appreciate some one else's point of view, and at the same time not fall prey to the "irrational exuberance" every new technology or technology application brings along with it. The key is to create effective strategies—the Decide phase of a strategic OODA.

Based on smashing together the work of insight and foresight, the resulting dialectic of thesis (this will happen), antithesis (that can't happen) and synthesis (ah, here is another way to achieve that outcome) the organization could articulate potential challenges and opportunities for itself and its customers in many ways, for example:

- Probability of occurrence
- Vertical segment (if applicable)
- Operational segment (if applicable)
- Geography
- Timeframe when the challenge would occur

- Possible technologies (available now or in the future) that could be used t
- Means to have checkpoints along the way that let us know if the probability
  and timing of the challenge is on track
- Which executives (CIO, CFO, others) within a company will be directly and indire
  ble for solving the challenge

Combine these and others with traditional strategic methods: SWOT (strengths, weakne opportunities and threats) analysis; McKinsey 7Ss; Porter's 5 forces; Ohmae's 3C's; Hammer's Im pact Value framework; Prahalad's Strategic Intent; and ADL's Matrix—whatever floats your boat. Then you can begin to set the boundaries and foci for innovation in your organization.

Two things should come out of this process (the DECIDE part of OODA)—and not necessarily a strategy per se. One thing that should come out of the process is a set of Strategic Goals, the strategic Action from OODA. These will drive the structural changes in the organization—new products, services, markets, and so on. The Strategic Goals support the *macro* innovation process and should drive three forms of innovation activity:

- *Plans of Record:* These are committed products and services that currently exist and are either being delivered or committed to customers. The influence of the strategic goals is to provide evolutionary guidance as to how they can begin to morph to fit the future, or perhaps more importantly, how they should begin to die out.
- *Plans of Intent:* These are the actions necessary to acquire or develop new capabilities within the organization so that when the time is right, they can be applied to new products and services not yet announced or delivered to customers.
- *Plans of Investigation:* These are research projects – technology, market, customer or combinations, often involving proof of concepts or the incubation of new offerings. There is just not enough information or confidence to invest into a plan of intent or a plan of record (actually building the product or service for customer use and scale) so the plans of investigation are macro innovation insurance policies to gain information to later decide one way or the other.

The other things to come out of this process are Reference Architectures. A significant amount of publicly available information has been published on the Internet regarding the definition of reference architectures. An architecture itself is the use of abstractions and models to simplify and communicate complex structures and processes to improve understanding and forecasting (see above). Although there is some divergence, common descriptive themes include:

- based on existing successful architectures, practices, and solutions;
- takes into account future needs and opportunities;
- written for a specific area of interest (domain);
- written to a certain level of abstraction (ranging from very specific to highly conceptual);

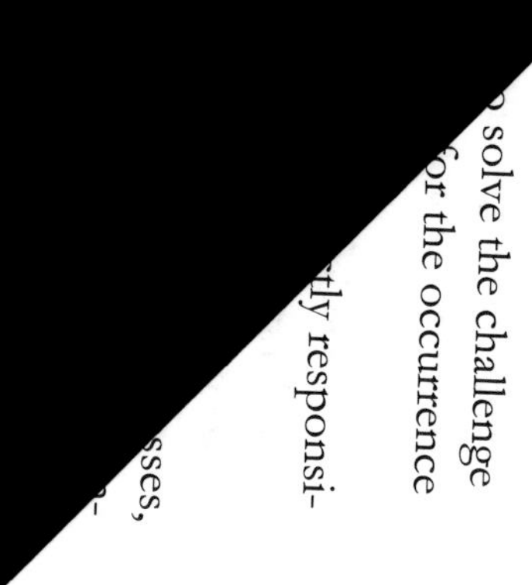

lity solutions; and
consistent communication.

ides the following criteria for a good reference architecture:
ıeterogeneous stakeholders (internal and external);
y the intended audience (defined and inclusive);
cted domain (identified goals);
stakeholders (metrics);
ty not sacrificed by desire to build consensus);
current); and

The purpose of the Reference Architecture is to enable at the *micro* level of OODA Action, numerous elements: a learning environment, creative thinking tools, design guidelines, engineering standards, research methods and tools, enabling work practices, an amenable culture, specific organizational structures, supportive management frameworks, numerous business processes, information systems, and the support infrastructure to allow and enable individual employees to look for, identify, investigate, communicate, suggest opportunities for innovation and then involve the organization as a whole for addressing them. In most cases three types of innovations and innovation processing have been observed:

- *Cross application innovation*: a technology, product, services, process or business model has been used successfully somewhere else, and it might be applicable to the problem, issue or opportunity at hand.
- *Continual improvement innovation:* an opportunity to do what is currently being done but in a faster, better or cheaper manner.
- *Problem-solving innovation:* using creative tools and techniques such as TRIZ, crowd sourcing and value challenges or open innovation fairs to address problems, issues or opportunities that there is no previous experience or track record (or success record) with.

Caution, however, is in order. Frameworks run the risk of suffering from the same mistakes organizations historically face with their strategy efforts. One difference and advantage is that the outcomes of frameworks (strategic goals and reference architectures) can be immediately translated into OODA *Actions*. But care should be taken to avoid the common pitfalls:[4]

- Failure to face the identified problems.
- Mistaking goals for strategy. Notice that the *macro* and *micro* actions are not financial—they are "do the work" activities so there is no "grow revenues at 20% with 20% margins," only what Action (OODA) is to be taken.

- Be as specific as possible. While innovation itself can be a fuzzy activity, you do not want the attention and focus on innovation to be fuzzy.
- Avoid fluff. The individuals in the organization who will be your source of innovation need to understand what you mean, so keep it simple and in plain language.
- You can't do everything, and your staff knows it—and your customers too, so be careful with the marketing and PR material The macro plans and micro support environment sets the playing field for what will be done, so either support your staff (resources and time) or don't include those activities you are not funding.

When thinking about innovation it is important to keep in mind that there are many forms of innovation. Innovation doesn't happen in a vacuum. People innovate in response to new challenges or new opportunities that emerge. Systematic innovation processes also typically begin with the identification of a dilemma that needs to be solved or a new opening for improvement. Changing the status quo requires proactive management, and the recognition of a problem or opportunity provides an incentive to try to do things differently, and serves as the driver for directing an innovation process.

- Innovation can be Disruptive, Breakthrough or Incremental to the work you do today.
- Innovation can be driven by business model change and the way your organization creates value. Southwest Airlines is the classic example of business model innovation. And Dell, for example, revolutionized the personal computer business model by collecting money before consumers' computers were assembled—creating a net positive working capital for seven to eight days. Such innovation addresses market and industry challenges, your plans and your strategy—often creating incubations, plans of investigation and plans of intent.
- Innovation can be driven by the industry or market or customers' changing needs and emerging opportunities impacting your offerings—changing your core products or services or changing the way you link or "bundle" offerings to create greater value. Microsoft, for instance, links a variety of its products, such as Word, PowerPoint, and Excel, into a more valuable "Office" bundle.
- Innovation can be driven by specific customer's immediate, ongoing or short term needs – continuous improvement through six sigma or constraint theory improvement, relative innovation taking ideas from one part of the business, market, industry or from far afield and applying it innovatively, or addressing problems that are not amenable to your existing ways of creating value. These innovations begin to change your (and your customers') supporting services in which you deliver a service beyond or around the core offerings; your core process and the way you create and deliver value and respond to customers; and your enabling processes (likely to increasingly be outsourced) that support your core business model, products and services.

Takeaway

Recognizing innovation as a systematic business process is far more important than just creating "an innovation," for if a company is to lead, it must set the "pace of innovation." To become a serial innovator, a company will need to view innovation as an ongoing *business process* that spans all the dimensions of business innovation.

Most companies equate innovation with the development of new products. But creating new products is only one of ten types of innovation, and on its own, it provides the least return.

Doblin, the renowned design firm started by Jay Doblin of Chicago's IIT Institute of Design, works with clients using a framework of Ten Types of Innovation that can help identify new opportunities in *finance, process, offering* and *delivery*. Companies that can simultaneously innovate across multiple innovation types will develop offerings that are more difficult to copy and that generate higher returns.

Innovation Category	*Innovation Type*	*Description of type*	*Business example*
Finance	1 Business model	How you make money	Dell revolutionized the personal computer business model by collecting money before the consumer's PC was even assembled and shipped (resulting in net positive working capital of seven to eight days).
	2 Networks and alliances	How you join forces with other companies for mutual benefit	Consumer goods company Sara Lee realized that its core competencies were in consumer insight, brand management, marketing and distribution. Thus it divested itself of a majority of its mfg. operations and formed alliances with mfg. and supply chain partners.
Process	3 Enabling process	How you support the company's core processes and workers	Starbucks can deliver its profitable store/coffee experience to customers because it offers better-than-market compensation and employment benefits to its store workers--usually part time, educated, professional, and responsive people.
	4 Core processes	How you create and add value to your offerings	Wal-Mart continues to grow profitably through core process innovations such as real-time inventory management systems, aggressive volume/ pricing/delivery contracts with merchandise providers, and systems that give store managers the ability to identify changing buyer behaviors in and respond quickly with new pricing and merchandising configurations.

Offerings	5 Product perform-ance	How you design your core offer-ings	The VW Beetle (in its original and its newest form) took the market by storm, combining multiple dimensions of product performance.
	6 Product system	How you link and/or provide a platform for multi-ple products.	Microsoft Office "bundles a variety of specific products (Word, Excel, PowerPoint, etc.) into a system designed to deliver productivity in the workplace.
	7 Service	How you provide value to custom-ers and consum-ers beyond and around your prod-ucts	An international flight on any airlines will get you to your intended designation. A flight on Singapore Airlines, however, nearly makes you forget that you are flying at all, with the most attentive, respectful, and pampering pre-flight, in-flight and post-services you can imagine.
Delivery	8 Channel	How you get your offerings to mar-ket	Legal problems aside, Martha Stewart has developed such a deep understanding of her customers that she knows just where to be (stores, TV shows, magazines, online, etc.) to drive huge sales volumes from a relatively small set of "home living" educational and product offer-ings.
	9 Brand	How you commu-nicate your offer-ings	Absolut conquered the vodka category on the strength of a brilliant "theme and variations" advertising concept, strong bottle and packaging design, and a whiff of Nor-dic authenticity.
	10 Customer experi-ence	How your custom-ers feel when they interact with your company and its offerings	Harley Davidson has created a worldwide community of millions of customers, many of whom would describe "being a Harley Davidson owner" as a part of how they fundamentally see, think, and feel about themselves.

Doblin's Ten Types of Innovation

References

[1] 12 February 2010, ID Number: G00174076 © 2010 Gartner Five Myths of Innovation, 2010 Update , Carol Rozwell, Kathy Harris

[2] http://computerperformance.co.uk/ezine/BestPractice/BestPractice64.htm

[3] http://www.thefreedictionary.com/blinkered

[4] www.mckinseyquarterly.com/Strategy/Strategic_Thinking/ The_perils_of_bad_strategy_2826

7. Open Innovation: Cast a Wide Net

The Tale of Stone Soup

Once upon a time, in a small village in a sad war-torn land, there was a great famine in which people jealously hoarded whatever food they could find, hiding it even from friends and neighbors. One day a wandering soldier came into the village and asked to stay for the night. "There's not a bite to eat in the whole province," he was told. "Better keep moving on."

"Oh, I have everything I need," he said. "In fact, I was thinking of making some stone soup to share with all of you." He pulled an iron cauldron from his wagon, filled it with water, and built a fire under it. Then, with great ceremony, he picked up a smooth stone from the ground and dropped it into the water.

Hearing a rumor of food, many villagers gathered around or watched from their windows. As the soldier sniffed the broth and licked his lips in anticipation, hunger began to overcome their doubts.

"Umm," the soldier said to himself rather loudly, "I do like a tasty stone soup. Of course, stone soup with cabbage – that's hard to beat."

Soon a villager approached hesitantly, holding a cabbage he'd retrieved from its hiding place, and added it to the pot. "Great!" yelled the soldier. "You know, I once had stone soup with cabbage and a bit of salt beef as well, and it was fit for a king."

The village butcher managed to deliver up some salt beef. And so it went, as one villager after another offered up potatoes, onions, carrots, mushrooms, and so on, until there was indeed a delicious meal for all.

The moral of the tale is simple. By collaborating and co-contributing the villagers could all enjoy a far tastier soup than each might have done alone—the power of *we*.

Not only is the Cloud where work gets done, it's also where innovation happens. "Modern economies," *The Economist* recently noted, "are not built with capital or labor as much as by ideas." But, as we've said repeatedly, those ideas are for naught without the capability to execute on them!

The brave new world of widely distributed knowledge in the Cloud has led to the business proposition of "open innovation." No longer can companies win the innovation arms race from the inside-out (internal R&D). They should, instead, buy or license innovations (e.g., patents, processes, inventions, etc.) from external knowledge sources, turning the table to outside-in innovation. Leading companies are turning to Social Networks to tap new sources of business intelligence and creativity that can lead to innovation. Don't confuse open innovation with open source, free software. Open innovation is all about the money to be made.

Open innovation models stress the importance of using a broad range of knowledge sources for a firm's innovation and invention activities, including customers, rivals, academics, and firms in unrelated industries while simultaneously using creative methods to exploit a firm's intellectual property (IP).

With the advent of easy-to-use "Consumer IT" or Web 2.0 usage of the Internet, social networks are changing the ways we live, learn, collaborate, work, consume and play. These huge changes in society also disrupt the way we design and manage our organizations and our value chains. Social networks have major consequences for business, even to the point of what it means to be a business, for it's no longer inside-out supply push, it's outside-in demand pull.

The customer is no longer king; the customer is now a dictator, and social networks are where customers get the information they need to make purchasing decisions. Your fancy Web site is the *last* place customers go.

Furthermore, inside your company, business units are increasingly turning to Cloud service providers to obtain the resources they need to get work done, collaborate and maintain relationships with customers. With credit card in hand, those resources are delivered to business units on demand, and often internal IT is the *last* place business units go for these resources.

The most important benefit of open innovation to companies is that it provides a larger base of ideas and technologies. Companies look at open innovation as a close collaboration with external partners—customers, consumers, researchers or other people that may have an input to the future of their company. The main motives for joining forces between companies is to seize new business opportunities, to share risks, to pool complementary resources and to realize synergies. Companies recognize open innovation as a strategic tool to explore new growth opportunities at a lower risk. Open technology sourcing offers companies higher flexibility and responsiveness without necessarily incurring huge costs.

Open innovation is more about increasing R&D options than about replacing existing ones. The external technological collaboration is complementary to internal R&D investments. An OECD study of 59 companies in a dozen countries found that almost three-quarters of them devoted the bulk of their R&D budget – 80% or more – to in-house R&D activities. At the same time most companies are actively involved in open innovation practices: 51% of the companies allocate up to 5% of their R&D budgets to research in other companies, while 31% allocate more than 10% outside.

Today there's much written on the co-creation of value with personalization and mass-customization of products and services. The *"co-creation of innovation"* is the new Stone Soup for the 21st Century.

The Gold Standard

Rob McEwen is the Founder and former Chairman and CEO of Goldcorp. He stunned the gold mining industry by sharing his company's proprietary geological data so that people all over the world could do the gold prospecting for him.

It's the year 2000 and the CEO of Goldcorp Inc., a gold producer headquartered in Vancouver, Canada was concerned about an underperforming mine in Ontario. The company's Red Lake mine was only producing a relatively small 50,000 ounces of gold a year at a high cost of $360 an ounce. The main deposits were deeper underground, but his company's geologists were not sure of the exact location of the precious metal

So he triggered a new gold rush by issuing an extraordinary challenge. He put all his company's geological data (which went back as far as 1948) into a file and shared it with the whole world. McEwen hoped that outside experts would tell him where to find the next six million ounces of gold. In return he offered $575,000 in prizes to the participants with the best methods.

Unsurprisingly his colleagues were skeptical, particularly as it was such a risky venture. The company was, after all, giving over its proprietary data as well as admitting to the industry that they were unable to find these elusive gold deposits.

The Goldcorp Challenge was launched in March 2000 and 400 megabytes worth of data about the 55,000 acre site was placed on the company's Web site. Everything that the company knew about the Red Lake mine was a mouse click away. Word spread fast around the Internet and within a few weeks submissions came in from all over the world as more than 1,000 virtual prospectors chewed over the data. In all more than 110 sites were identified and 50% of these were previously unknown to the company. Of these new targets, more than 80% yielded significant gold reserves. McEwen believes that this collaborative process cut two, maybe three years off the company's exploration time. And the worth of this gold has so far exceeded $6 billion in value. The prize money was only a little over half a million dollars, so it was a fantastic value for money investment, and much cheaper than continuing with unproductive exploratory drilling.[1]

GE's Ecomagination

In July, 2010, GE's Chairman and CEO, Jeff Immelt announced a unique crowd sourcing project, wherein entrepreneurs, technologists and start-ups were encouraged to submit ideas that could be likely to revolutionize GE's next generation electric grid. The challenge started with a $200 million capital pledge by GE and its partners—venture capital firms Emerald Technology Ventures, Foundation Capital, Kleiner Perkins Caufield & Byer, and RockPort Capital.

Considering the speed at which smart grid innovation is developing it would have been difficult for any one company's R & D department to tackle this project on its own. Thus, GE being a global player envisaged this Ecomagination challenge as a means of inviting ideas from everywhere and then combining and deploying them on a global scale. The objective was to generate ideas that are compatible with ecological sustainability or green technology in the fields of renewable energy, grid deployments, smart homes and buildings.

The challenge lasted 10 weeks. GE's partners, as well as independent advisors and analysts helped GE to sort through the entries. The general public was also allowed to vote. A $100,000 award was available for entries demonstrating "outstanding entrepreneurship and innovation." In addition to this, commercial opportunities in the form of distribution relationships, R&D validation, and partnerships that might help smaller start-ups get products to market more quickly were also announced. Given the market clout of GE, the promising innovations are likely to see the light of the market very quickly. Immelt said "We want GE to be a good aggregator for innovators and for our customers. In GE, we have lots of great researchers but we're not going to invent every idea."

GE is uniquely placed in the sense that it has tremendous marketing and distribution clout to be able to put these new innovative technologies in the market very fast. Immelt said Ecomagination would release approximately 30 products over the next 24 months.

Through 2015, GE plans to spend $10 billion on projects to back its Ecomagination philosophy for encouraging green technology and sustainability R&D. The technologies and products that are expected to come out of the initiative include battery chemistries, wind turbines, solar and desalination technologies, energy-efficient appliances, new aircraft and locomotive engines.

GE launched the Ecomagination challenge again in January 2011, this time with a specific objective of soliciting ideas for improving energy efficiency in homes using new technologies. In this round, it once again awarded 10 companies for new concepts, like SunRun's solar home power services, and five Innovation Award winners for start-up organizations like Pythagoras.

SunRun offers homeowners its solar power service, a lease-type agreement that allows homeowners to reap the benefits of solar systems on their homes with little to no upfront cost. Pythagoras, a California-based company, with offices in Israel and China, has developed what it calls the PVGU (photovoltaic insulating glass unit).

Dell's IdeaStorm

Dell's IdeaStorm was launched in 2007 to give a direct voice to Dell's customers and an avenue to have online "brainstorm" sessions to allow customers to share ideas and collaborate with one another and Dell. The IdeaStorm concept also plays a critical role in Dell's efforts to build the company's partner network of IT consultants, systems integrators and value-added resellers.

As of mid-2011, IdeaStorm had crossed the 15,726 idea mark and implemented over 400 ideas. In addition to the open discussions at the IdeaStorm site, Dell posts a specific topic and asks customers to submit ideas. These Storm Sessions are only open for a limited time, making the sessions targeted, relevant and time bound.

By registering at the site users can add articles and the community can comment, promote, and demote them. Promoting an article increases its score, which allows Dell to rank suggestions and requests by their importance to the site's users. Articles can also be demoted, and a "vote half life" system is used to remove the ideas that are no longer receiving votes from popular ideas page.

A page is also maintained to demonstrate how Dell is acting on the suggestions. This page is only changed when the status of an idea changes to *Implemented.* The ideas range from everyday issues like requests that Dell's technical support lines are operated in the specific countries in which Dell computers are sold, thus avoiding unnecessary breakdowns in communication.

There are also the more technical requests. The site hosts discussions wherein ideas on processor specs, aspect ratios and all types of technology improvements are thrown about for brainstorming. An important example of this is when readers flooded IdeaStorm with suggestions and proposals when Dell began exploring the desktop Linux market in early 2007. Ultimately, IdeaStorm played a critical role in Dell's decision to pre-load Ubuntu Linux on selected laptops and desktops.

IdeaStorm was followed by Employee Storm. Employee Storm consisted of employees writing original articles and then answering questions and responses to them. And in reality answering was more important than posting articles. Statistical success can be measured by the fact that Employee Storm has received over 4,100 ideas with 225,000 votes and 18,500 comments. Such achievements have surpassed the initial purpose of the project and have resulted in fast evolution of the work culture. Employees now like the sharing of ideas through blogs and find information easily through transparent means.

Achievements of Employee Storm have surpassed the initial purpose of the project and have resulted in fast evolution of the work culture. Employees now like the sharing of ideas through blogs and find information easily through transparent means.

Procter & Gamble's Connect and Develop

Procter & Gamble used to operate one of the greatest R&D operations in corporate history. But as the company grew to a $70 billion enterprise, the global innovation model it devised in the 1980s was not up to the task. In 2001, Larry Huston, Procter & Gamble's then Vice President of R&D Innovation and Knowledge, was given a lofty goal by CEO A.G. Lafley—source 50% of the company's innovation externally. To help harvest the requested crop of external ideas, Huston developed a program to tap into a variety of networks consisting of the smartest scientists and business minds around. In the academic world, they called it "Open Innovation." At P&G, it became known as "Connect and Develop."

All this not only required operational changes but also mind set changes to make the boundary porous between P&G's R&D department and the thousands of innovators on the outside. Some of the achievements made possible through Connect and Develop include:

- P&G's productivity has increased by 60%
- The innovation rate has doubled
- The cost of innovation has fallen
- R&D investment as a % of sales fell from 4.8% to 3.4%
- Five years after the company's stock collapse, the shares have doubled in value.

Quite unlike outsourcing, P&G's Connect and Develop scouts for individuals and companies with innovative ideas and then brings them inside to enhance and capitalize P&G's production and marketing capabilities. Olay Regenerist, Swiffer Dusters and Crest SpinBrush are some of the successes of this innovative model. It was very important to have a clear plan as to what products or ideas to select from. Foremost is that they already have a proven track record or acceptance rate among the consumers. That track record allows P&G to enhance its own technologies and then promote them successfully with its widely developed marketing and distribution capabilities. This synergistic relationship has proved to be mutually beneficial.

In order to undertake the Connect and Develop activities, P&G maintains network connections with industry and university researchers. They work from six main hubs: China, India, Japan, Western Europe, Latin America and the U.S. They have so far identified thousands of products, product ideas and promising technologies. P&G soon saw that their supplier networks are endowed with an R&D staff of 50,000 who can be potential innovators, and immediately set about collaborating with them by sharing their technology briefs with suppliers. In a unique example of "co-creation" the suppliers' researchers work in P&G's labs and at times P&G researchers work in the suppliers' labs. P&G also taps open innovation resources including the NineSigma open innovation network, InnoCentive (founded by Eli Lilly), YourEncore (retired scientists and engineers) and Yet2.com, an intellectual property exchange.

In 2008, Huston left P&G to found 4Inno.com. Huston wisely advises that it has become imperative to adapt to the open innovation model in order to survive the 21st century challenges and competitions. He warns that adhering to the old invent-it-yourself model ultimately leads only to diminishing returns.

Watch Larry Huston: tinyurl.com/75g3c5s

tinyurl.com/7dkev5z

Don't miss a down-to-earth conversation with Lafley:

tinyurl.com/77ueuso

IBM's Global Innovation Jams

In July 2006, IBM kicked off a defining moment in collaborative innovation: arguably the largest online brainstorming session ever held. The event was the IBM InnovationJam and it attracted more than 150,000 participants from 104 countries and 67 different companies over the course of two 72-hour sessions.

They were drawn by IBM CEO Sam Palmisano's objective to invest up to US$100 million to develop and bring to market the best ideas from the event. "We opened up our labs and said to the world, 'Here are our crown jewels, have a look at them,'" said Palmisano.

Their dialog resulted in tens of thousands of creative and far-reaching ideas, many of which are already having an impact on business and society today.

IBM managers then used automation to winnow the 37,000 ideas they received down to 300 well-defined ideas. Finally, more than 50 employees spent a week at IBM's Watson Research Center in New York further combining and trimming these top ideas down to 30. And now the company is spending $100 million to develop the ideas that came from the Jam.

"Collaborative innovation models require you to trust the creativity and intelligence of your employees, your clients and other members of your innovation network," said Palmisano. "The Jam—and programs like it—are greatly accelerating our ability to innovate in meaningful ways for business and society."

May you Jam in interesting times! InnovationJam 2008 focused on "The Enterprise of the Future"—based on interviews with more than 1,100 CEOs. In the midst of a subprime debacle, housing market collapse, investment bank failures and plenty of ominous developments still gathering on the horizon, what initially appeared to be the worst of times for a global brainstorming session on how to build the 21st century enterprise was quickly judged to be the best time possible by the tens of thousands of 2008 participants. What better time for effecting real change than in the full fury of a global burning platform?

Events around the world since the four and a half-day InnovationJam continue to underscore the uncertainty and depth of the challenges ahead. They also validate the importance of the fundamental issues identified by Jammers: enabling and benefiting from greater transparency; increasing efficiency in all aspects of a company's operations; and expanding the idea of business stewardship to economic, social and environmental sustainability.

IBM's Innovation Jams

To repeat, don't confuse open innovation
with open source, free software.
Open innovation is all about the money to be made.

Weak Ties and The New Polymaths

Does your company have a Chief Polymath Officer? Don't worry. We're not suggesting that your company needs a Chief *Math* Officer. Polymath is the Greek word for a Renaissance person like Leonardo Da Vinci or Ben Franklin who excel in many disciplines. In *The New Polymath*, former Gartner analyst, Vinnie Mirchandani, describes the enterprise that has learned to amalgamate multiple strands of technology (infotech, cleantech, healthtech, nanotech, biotech) to create *compound* new products and to innovate internal processes. He details 11 building blocks these New Polymaths are utilizing, from cloud computing to sustainability to Social Networks.

While your own innovation team may be good at tweaking, it also needs big, breakthrough thinking. The New Polymaths are not just amalgamating. They are learning new disciplines all the time and leveraging the state of the art in multiple technologies. Mirchandani organizes those technology components using a R-E-N-A-I-S-S-A-N-C-E framework where each letter represents a building block for the New Polymath to leverage. I, for example, covers interfaces other than keyboard/mouse interfaces. The second A covers Analytics (predictive, Web, data visualization) that are way more useful than old business intelligence (BI). C stands for cloud computing. The first N covers Networks as in innovations in telecom, the second N also covers Networks as in the human network – communities, crowds and collaboration. The S's cover Sustainability for cleantech trends and Singularity for healthtech trends. One of the Es covers Ethical issues that are proliferating as compound innovations are brought to market.

This multidisciplinary approach to Polymath-kinds innovation is really about reinforcing the "strength of weak ties." Many people attribute great creative achievement to *genius*, whatever that really means. Genius is often perceived as the thinking that goes on inside the heads of people like Edison and Newton, or in modern times the likes of CEOs such as GE's Jack Welch or Apple's Steve Jobs. Unfortunately, mainstream psychology hasn't been able to throw much light on what genius actually is. However a new paradigm is emerging that directly challenges a major assumption of most current theories of creative thinking, namely that the intelligence that drives it is located exclusively inside the head (the Mind-Inside-the-Head [MITH] model).

Leading philosophers and mind/brain scientists like Daniel Dennett and Andy Clark argue that contrary to both the academic and commonsense view, the mind extends out into the world. As Clark succinctly puts it, "We use intelligence to structure our environment so we can succeed with less intelligence. Our brains make the world smart so we ourselves can be dumb in peace. Once we recognize the existence of external sources of intelligence, we can begin to conceive of creative thinking in a radically new way." "The strength of weak ties" is the term coined in 1973 by Mark Granovetter in what many now regard as a seminal work defining "social" collaboration. Strong ties represent the people you are closest to – coworkers, nuclear family, friends, supply-chain partners and so on. Weak ties are connections to people that you may occasionally come across – a friend's friend or online communities that share special interests outside of your ordinary interests. Granovetter argues that strong social ties are good for exerting power, but because they contain a lot of redundant information they are almost useless for gaining fresh information, new perspectives and insights – the raw materials for innovation. In contrast, weak ties contain much less redundant information and are often more important for gaining fresh information, connecting new dots and thinking outside your own box.

The human interactions with previous IT collaboration tools have focused on "strong ties," building the capabilities for relationships between known people around known topics in a manner that provides a structure to deliver value. This remains an important aspect of workflow, but if we consider the emerging Social Networks, then it becomes clear that something different, something unstructured, something unknown is needed to bring new value to the table.

As reported by Patty Azzarello, a former general manager at HP, "A large network of 'weak connections' is more valuable than a small network of close connections. And it is not just a matter of the numbers. The people you are close to are not always very useful to help you because they tend to be in the same environments, know the same people, and think similarly to you. Whereas your 'weak connections' have access to different stuff, people, places and things."

It's the responsibility of the next-generation leadership to establish the capabilities needed for "that something different" for facilitating "the strength of weak ties." To do that the next-generation leaders and knowledge workers must cross disciplines and heed the advice of Ram Charan, the guru of management gurus, who has no home, clocks 500,000 air miles a year, and pulls down $20,000 a day talking simple talk to executives across the globe. He lives nowhere, and goes everywhere:

- "Continually search for the new and different.
- Sift, sort, and select.
- Hit many singles and doubles, not just home runs.
- Put yourself in positions where you can expand your view.
- Listen, read, become a keen observer.
- Connect the dots to make sense of it all.
- Keep your perceptual lenses open, act on your own curiosity.
- Be proactive in shaping your view of which way things are going."

The spoils belong to those who can act ahead of others – because they see things ahead of others. Much of what Charan and other leaders do is to help business people bring common-sense clarity to an otherwise complex business world. It's all too easy to lose such clarity in the day-to-day grind of the real world of work. But considering all the big business issues—innovation, globalization, profitable growth, increased efficiency, market planning, social responsibility and increased productivity—a common thread for achieving clarity is to look from the outside-in.

It's wrong to think the dramatic and unexpected changes we are experiencing in today's world will somehow fade away and we'll return to business as usual—the old normal. There's no going back to the good ole days. Innovate or fade away. It's time to tap the power of weak ties as a catalyst for innovation. And that's the real job of the next-generation CEO.

Some Open Innovation Issues

We identify three fundamental challenges for firms in applying the concept of open innovation: finding creative ways to exploit internal innovation, incorporating external innovation into internal development, and motivating outsiders to supply an ongoing stream of external innovations. This latter challenge involves a paradox. Why would firms spend money on R&D efforts if the results of these efforts are available to rival firms? Lessons can be learned from the world of commercial open source software. Firms involved in open source software often make investments that will be shared with real and potential rivals.

The term "open innovation" does not refer to free knowledge or technology. While "open source" refers to royalty-free technologies, "open innovation" refers to the collaborative methods applied, and may still imply the (significant) payment of license fees between companies for intellectual property.

Open innovation itself benefits from the same innovation architecture described in Chapter 6. By bounding the "challenges" issued in open innovation via focused constraints and operating parameters, the probability of success increases significantly. The lessons learned from open source and the development of innovation architectures include:

- Carefully craft the problem statement so it's in plain language and easily understood.
- Experiment in-house first – share the results, otherwise everyone is starting from scratch.
- Offer incentives—and they don't have to be financial.
- Set a time limit.
- Create a transparent evaluation process—people will only participate if they think the process is "fair."
- Say how you plan to complement the outcome—demonstrate the value of participation!
- Foment organizational transformation from the outside-in—invite lead users to create derivative works out of your intellectual property, to share their creative ideas with one another, and to build their own "solutions" (gadgets, mash ups, applications, etc.) leveraging your company's branded IP.
- Host co-design sessions with lead users. Invite creative professionals from multiple disciplines to creative workshops to co-design new concepts. Select the best of these concepts to carry further, sponsor, and commercialize.
- Encourage customers to contribute ideas and content, to pose and solve problems, and to interact with one another in public online community spaces.
- Encourage your own employees to leverage customer-contributed content, ideas, and deliverables.
- Provide tools like high-level programming languages and toolkits to promote lead user innovation—and offer training on those tools; make sure that each training class produces real deliverables or at least prototypes of new designs.
- Get all your stakeholders aligned around customers' desired outcomes. Provide integrated, cross-disciplinary services and support to help customers reach their outcomes; enable customer-led, individually-optimized service delivery.
- Create expert networks and link customers to networks of experts. Make it easy for people to find the experts they need.
- Empower local community-based problem solving. Provide support and structure to enable community members to collaborate to solve common problems.
- Provide tools to end users/customers to manage their own complex situations (health, projects, etc.) rather than trying to do things for them.
- Provide electronic design tools to interested end users/customers to design their own products and to design your company's products in open design communities. Encourage customer de-

signers to critique and vote on each others' work.
- Harvest user-generated ideas from across the Internet. Pull it all together and look at the patterns of needs and solutions.

Open Innovation Platforms

Here are concrete examples of open innovation platforms and initiatives listed at OpenInnovators.net: [2]

Intermediary Platforms

Research & Development Platforms

- Innocentive - open innovation problem solving
- TekScout - crowdsourcing R&D solutions
- IdeaConnection - idea marketplace & problem solving
- Yet2.com - IP market place
- PRESANS (beta) - connect and solve R&D problems
- Hypios - online problem solving
- Innoget - research intermediary platform
- One Billion Minds - online (social) challenges
- NineSigma - technology problem solving

Marketing, Design & Idea platforms

- RedesignMe - community co-creation
- Atizo - open innovation market place
- Innovation Exchange - open innovation market place
- Ideaken - collaborative crowdsourcing
- Idea Bounty - crowdsourcing ideas
- Guerra Creativa - anything from logos to websites
- Brand Tags - tagging brands
- Battle of concepts - student challenges
- Brainrack - student challenges
- CrowdSPRING - creative designs
- BootB.com - custom ideas for any creative need
- Myoo Create - environmental and social challenges
- 12designer - marketplace for creative solutions
- LeadVine - crowdsourcing lead generation
- 99designs - pioneer in design crowdsourcing
- Edge Amsterdam - elite sourcing platform
- OpenIDEO - collaborative design platform

- Challenge.gov - solutions for government problems
- eYeka - the co-creation community

Collective Intelligence & Prediction platforms

- Inkling Markets - the wisdom of crowds for forecasting
- Intrade - global prediction markets
- NewsFutures - collective intelligence markets
- Ushahidi - crowdsourcing crisis information
- Kaggle - data mining and forecasting
- We Are Hunted - the online music chart
- ESP Game/Google Image Labeler - image labeling

HR & Freelancers platforms

- TopCoder - competition-based software
- Spudaroo - crowdsourcing copywriting
- HumanGrid - small online task solving
- ChumBonus - crowdsourcing recruitment
- Amazon Mechanical Turk - low-cost crowdsourcing

Open Innovation Software

- Spigit - idea management 2.0
- Imaginatik - collective intelligence software
- Napkin Labs - connect with consumers, experts
- Fellowforce - software/suggestion box 2.0

Intermediary Open Innovation Services

- Big Idea Group - innovation contests and idea hunts
- Idea Crossing - organize innovation quests
- Pharmalicensing - open innovation for the life sciences
- Chaordix - crowdsourcing engine for innovation
- DataStation - complete innovation platform

Creative Co-creation

- Spreadshirt - shirt community
- JuJups - personalized gifts
- Threadless - create and sell your t-shirts
- Naked&Angry - threadless for ties and wall coverings
- Cafepress - shop, create or sell what's on your mind
- Cazzle - create and sell products
- CreateMyTattoo - crowdsourced tattoo design
- Sellaband – crowdfunded bands

- Artistshare – fans funding new artists
- Quirky - community product development
- Jovoto - co-creation & mass collaboration
- Dream Heels - design your dream heels

Corporate Initiatives

Product Ideas Crowdsourcing

- Ideas Project - crowdsourcing platform by Nokia
- Fiat Mio - create a car
- Open Innovation Sara Lee - open innovation portal
- P&G Open Innovation Challenge
- Ideas4Unilever - corporate venturing
- BMW Customer Innovation Lab - in German
- LeadUsers.nl & Live Simplicity - Philips' crowdsourcing platforms
- Kraft - innovate with Kraft
- InnovationJam - IBM's idea generation project
- Dell IdeaStorm - external idea sourcing
- Vocalpoint - P&G's network for women
- Betavine - Vodafone's mobile app community
- My Startbucks Idea - shaping the future of Starbucks

Branding & Design Crowdsourcing

- Spreadshirt Logo Design Contest - design a new logo
- Gmail M-Velope - viral video competition
- LEGO Factory - LEGO co-creation tool
- Peugeot - Peugeot's design contest
- Muji - improving and suggesting new designs
- Electrolux Design Lab - competition for students
- Fluevog - open shoe design
- LEGO Mindstorms - open source robots
- BurdaStyle - open source sewing
- GoldCorp - the famous GoldCorp Challenge

Peer Production & P2P

- CrowdSpirit - product development 2.0
- Funding Circle - p2p lending
- Linux - open source software
- Wikipedia - peer produced encyclopaedia
- Yahoo Answers - crowdsourced Q&A

- A Swarm of Angels - creating a £1 million film

Public Crowdsourcing

- iBridge Network - platform for university innovation (iBridge)
- Science Commons - generic license agreements
- Picnic Green Challenge - ideas to save the planet
- Eureke medical - medical open innovation platform
- German Catholic Church - adopts open innovation
- Fold it - solve puzzles for science
- Ideas Campaign - citizen ideas in Ireland
- Galaxy Zoo - discovering the universe

Some of the above look like collections of amateurs. Maybe so. But just listen to a researcher at the London think tank Demos, Charles Leadbeater, on the rise of the *amateur professional*:

tinyurl.com/oqzc84

Since innovation has increasingly become the basis for the competitive advantage of companies, the growing number of interactions with external parties such as customers, suppliers, universities, etc., has important repercussions for the protection and safeguarding of intellectual assets and intellectual property – patents, trademarks, trade secrets, etc. Open innovation may increase the risk of leakage of proprietary knowledge and involuntary spillovers. Other potential disadvantages are the extra costs of managing co-operation with external partners, the loss of control, the adverse impact on the flexibility of the company, the dependence and possible over-dependence on external partners, and the potential opportunistic behavior of partners.

But let's be realistic. The days of innovating in isolation are over. No one company can be expected to know all the answers. The largest patent holder in the world (IBM) altered its corporate policy on the management of patents in 2006 to enable open innovation. IBM also released its business methods patents to the public domain and pledged to do so with future patents as well.

IBM and others have recognized two things. First, no single firm is able to develop all the technology, methods and processes it needs internally. Second is that the products and services they produce need to work well with those produced by other firms, even including direct competitors (coopetition) and firms with very different business models. The risk management question now becomes, is the loss of value greater in losing IP or in losing future innovation?

In this setting, it is essential that firms develop new ways to ensure that they retain some of the profits accruing from "open innovation" projects and development—things like royalty free access to collaborative outcomes and extensive cross licensing. This in turn has led to what Arora, Fosfuri, and Gambardella have called the "markets for technology."[3]

Takeaway

Intellectual property theft is typically identified as the most important risk of global innovation networks. Unique knowledge may be revealed to external partners who could later become competitors or who could make better use of the fruits of the know-how. Working closely with external partners can create uncertainty about how the benefit of technology collaboration is to be appropriated.

Small and medium enterprises (SMEs) may be confronted with larger risks in collaborating with larger companies because of their typically smaller resources and limited expertise in intellectual property rights (IPR) issues. But this too is changing as you can learn from former Reuter's CIO, Ken Thompson.

Read the overview: mkpress.com/TNE
or tinyurl.com/7qgjtxj

Investing in people and fostering cross-functionality and mobility and a "culture of innovation" is crucial, as open innovation implies that people must be able to work in networks and across borders, sectors and at the interface of converging technologies. It also requires openness to a geographically mobile workforce.

We hope we've piqued your interest in open innovation and want to point you to a useful reference for further study: *A Guide to Open Innovation and CrowdSourcing; Advice from Leading Experts.*

tinyurl.com/74fjc9m

Watch M.I.T.'s Thomas Malone on crowdsourcing and "The Age of Hyperspecialization."

tinyurl.com/7ankblx

Related Article: *tinyurl.com/447xsu8*

References

---

[1] http://www.ideaconnection.com/open-innovation-success/Open-Innovation-Goldcorp-Challenge-00031.html

[2] http://www.openinnovators.net/list-open-innovation-crowdsourcing-examples/

[3] Arora, Ashish, Andrea Fosfuri, and Alfonso Gambardella. (2001). Markets for Technology. Cambridge, MA: MIT Press.

8. Innovation and Social Networks

Social BPM, Social CRM—Social *This* and *That*—are all the rage in the business world today. Yada Yada Yada about these topics that abound, thanks to bloggers, IT analysts and business writers. *Yawn.* Here's what should really have captured our attention:

Solicitation Number: RTB220610
Added: Jun 22, 2010 1:42 pm
Agency: Department of the Air Force
Office: Air Mobility Command
Location: 6th Contracting Squadron

0001- Online Persona Management Service. 50 User Licenses, 10 Personas per user. Software will allow 10 personas per user, replete with background , history, supporting details, and cyber presences that are technically, culturally and geographically consistent. Individual applications will enable an operator to exercise a number of different online personas from the same workstation and without fear of being discovered by sophisticated adversaries. Personas must be able to appear to originate in nearly any part of the world and can interact through conventional online services and social media platforms. The service includes a user friendly application environment to maximize the user's situational awareness by displaying real-time local information.

0002- Secure Virtual Private Network (VPN). 1 each. VPN provides the ability for users to daily and automatically obtain randomly selected IP addresses through which they can access the internet. The daily rotation of the user's IP address prevents compromise during observation of likely or targeted web sites or services, while hiding the existence of the operation. In addition, may provide traffic mixing, blending the user s traffic with traffic from multitudes of users from outside the organization. This traffic blending provides excellent cover and powerful deniability. Anonymizer Enterprise Chameleon or equal

0003- Static IP Address Management. 50 each. License protects the identity of government agencies and enterprise organizations. Enables organizations to manage their persistent online personas by assigning static IP addresses to each persona. Individuals can perform static impersonations, which allow them to look like the same person over time. Also allows organizations that frequent same site/service often to easily switch IP addresses to look like ordinary users as opposed to one organization. Anonymizer IP Mapper License or equal

0004- Virtual Private Servers, CONUS. 1 each. Provides CONUS or OCONUS points of presence locations that are setup for each customer based on the geographic area of operations the customer is operating within and which allow a customer's online persona(s) to appear to originate from. Ability to provide virtual private servers that are procured using commercial hosting centers around the world and which are established anonymously. Once procured, the geosite is incorporated into the network and integrated within the customers environment and ready for use by the customer. Unless specifically designated as shared, locations are dedicated for use by each customer and never shared among other customers. Anonymizer Annual Dedicated CONUS Light Geosite or equal

0005- Virtual Private Servers, OCONUS. 8 Each. Provides CONUS or OCONUS points of presence locations that are setup for each customer based on the geographic area of operations the customer is operating within and which allow a customer's online persona(s) to appear to originate from. Ability to provide virtual private servers that are procured using commercial hosting centers around the world and which are established anonymously. Once procured, the geosite is incorporated into the network and integrated within the customers environment and ready for use by the customer. Unless specifically designated as shared, locations are dedicated for use by each customer and never shared among other customers. Anonymizer Annual Dedicated OCONUS Light Geosite or equal

0006- Remote Access Secure Virtual Private Network. 1 each. Secure Operating Environment provides a reliable and protected computing environment from which to stage and conduct operations. Every session uses a clean Virtual Machine (VM) image. The solution is accessed through sets of Virtual Private Network (VPN) devices located at each Customer facility. The fully-managed VDI (Virtual Desktop Infrastructure) is an environment that allows users remote access from their desktop into a VM. Upon session termination, the VM is deleted and any virus, worm, or malicious software that the user inadvertently downloaded is destroyed. Anonymizer Virtual Desktop Infrastructure (VDI) Solution or equal

Contracting Office Address: 2606 Brown Pelican Ave., MacDill AFB, Florida 33621-5000 United States
Place of Performance: Performance will be at MacDill AFB, Kabul, Afghanistan and Baghdad, Iraq.

Are we on the cusp of social engineering armed with sock puppets, trolls and drones dispatched as a result of real-time predictive analytics to do our bidding in social networks? Science fiction? —not.

While "analytics" are becoming part and parcel of most BPM systems and are the hot topics of BPM experts such as Thomas Davenport, we need to look beyond—to Manufacturing Consent 2.0. What's hot about analytics is that instead of looking backward to analyze "what happened?" predictive analytics help executives answer "What's next?" and "What should we do about it?" Contemporary analytics is all about looking forward to make fact-based decisions about what *might happen* versus looking backward (as in that old-fashioned Business Intelligence stuff) to understand what *has happened.* But there's even more. Manufacturing Consent 2.0 serves as the engine, steering wheel and accelerator for going beyond predictive analytics and on to *action.* Manufacturing Consent 2.0 is about *manipulating, versus predicting, what happens next*, and what happens next takes place inside Social Networks around the globe.

Noam Chomsky's 1988 book, *Manufacturing Consent: The Political Economy of the Mass Media*, uses the propaganda model and posits that corporate-owned print, radio, and television news sources are businesses that distort news reportage—what types of news, which items, and how they are reported. To minimize financial danger, news media businesses editorially distort their reporting to favor government and corporate policies in order to stay in business. In short, Chomsky asserts that *propaganda in a democracy* replaces *violence in dictatorships* as a means of serving the elite and powerful. Today, extend the ideas from 1988's one-to-many publishing and broadcasting to today's many-to-many social media where the "truth" is *constructed* through two-way conversations and you have Manufacturing Consent 2.0.

As written in *Enterprise Cloud Computing,* "Social constructionism focuses on the 'truths' that are 'created' through the interactions of a group. Social constructs are the by-products of countless human interactions and choices rather than laws resulting from nature, as revealed in science. Social constructionism is about the categorical structure of reality—the ways individuals and groups participate in the creation of their *perceived* reality that ultimately becomes institutionalized and made into tradition. Socially constructed reality is an ongoing, dynamic process. Reality is reproduced by people acting on their interpretations. In short, people are driven by stories and those narratives lead to social constructionism."

A simple example can be seen in the "cult of Apple." There is no shortage of Apple fanatics who eat, think, and breathe Apple, devoted to all things Apple, Macintosh, iPod, iPad and iPhone. In fact, it is so well established that Samsung released an ad mocking it while this book was being written:

tinyurl.com/bldlmpf

While a master at both innovation and the manufacturing of consent, Apple may actually be upping the game with the release of Siri. With Siri, its voice activated assistant for the iPhone 4S, Apple is moving beyond simple voice control and into creating a *personality*, and perhaps a "*trusted friend*." For the first time, our phones are emoting to us, and we're being asked to do the same to them. There is a blog from 23rd October 2011 by Matthew Panzarino called "Siri, Mickey Mouse and Apple's cult of personality," where he very effectively compares Siri as it is to Mickey as he was (just a foil or tool to get laughs) in the beginning, then how could Siri evolve to the role Mickey has now – the trusted spokesperson for Disney, and a social exemplar for adults and children (ignoring the early films of course). The difference, Mickey has a persona (created by animators) and Siri does not. We now see that can change. [1]

Also, as we have mentioned several times in this book, maintaining context – location, previous requests, proximity – are all key components of future innovative services, and all innovations will have service components. Siri not only gives you the ability to speak as you would to a human and not have to reconfigure your speech patterns and thought processes to speak to a computer, but like a true friend or assistant knows where you are, what you are interested in and can make a reasonable guess as to your intent. That can manufacture a lot of consent and influence.

As an aside, remember the military contract at the beginning of this chapter? Siri has its roots in a project called Cognitive Assistant that Learns and Organizes, which was funded in part by the U.S. Department of Defence. And let's not forget Watson, that IBM creation that swept the world when it won against the two very best players of Jeopardy!—all voice driven.

Listening Posts

Because social networks are "where they are talking about your company, your products and your services," step one in the brave new world of social constructionism is listening to what people are saying about you. Traditional focus groups, surveys and market research studies won't cut it in today's connected world. Today, we need to be actively listening all of the time to our customers, influencers, competitors and communities of people that can impact our businesses.

Listening is a critical part of the OBSERVE and ORIENT cycle of OODA for innovation and the real-time FORESIGHT component of the Innovation Architecture. Social media have established a constant flow, actually a torrent, of conversations. So if you want to have the very best analytics possible, then you first have to have the means to *listen* to the flood of chatter and jabber.

PepsiCo recently created the Gatorade Mission Control Center inside of its Chicago headquarters, a room that sits in the middle of the marketing department and could best be thought of as a war room for monitoring the brand in real-time across social media. The room features six big monitors with five seats for Gatorade's marketing team to track a number of data visualizations and dashboards—also available to employees on their desktops. Mission Control measures social media conversations across a variety of topics and shows how hot those conversations are in real time.

The system also runs detailed sentiment analysis around key topics and product launches.

Meanwhile, Dell's Ground Control tracks more than 22,000 daily topic posts related to Dell, as well as mentions of Dell on Twitter. The information is sliced and diced based on topics and subjects of conversation, sentiment, share of voice, geography and trends. Such systems also must be on the lookout for competitors as well as customers, for competitors are likely to post *trolls* raising negative points.

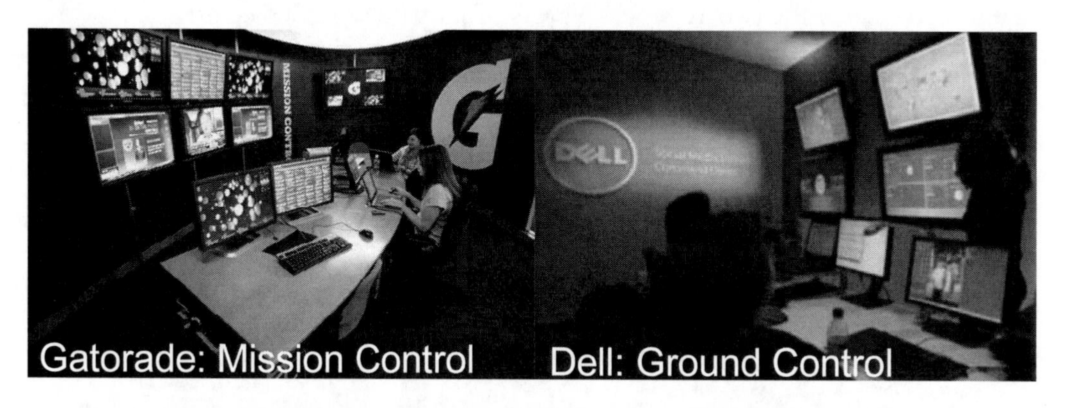

Gatorade: Mission Control Dell: Ground Control

The investment arms of the CIA and Google are both backing a company that monitors the Web in real time to predict the future.

The company, Recorded Future, scours thousands of websites, blogs and Twitter accounts to find the relationships between people, organizations, actions and incidents, both present and still-to-come. The company says its temporal analytics engine "goes beyond search" by "looking at the 'invisible links' between documents that talk about the same, or related, entities and events." The idea is to figure out for each incident who was involved, where it happened and when it might go down. Recorded Future then plots that chatter, showing online *momentum* for a given event. "The cool thing is, you can actually predict the curve, in many cases," says the CEO Christopher Ahlberg, a former Swedish Army Ranger with a PhD in computer science.

Then again, the CIA and the wider intelligence community is putting cash into Visible Technologies, a software firm that specializes in monitoring social media. Visible crawls over half a million Web 2.0 sites a day, scraping more than a million posts and conversations. Then Visible "scores" each post, labeling it as positive or negative, mixed or neutral. It examines how influential a conversation or an author is, trying to determine who really matters.

As this book was being written, CBS released a new television show "Person of Interest" based upon a government developed system used to predict terror attacks and capable of "ballparking" individual acts of crime (the computer can say something bad will happen involving a specific someone, but can't say what will happen or whether the person is the bad guy or the good guy). We will stick by our assertion that science fiction should be part of your "foresight portfolio."

In-Q-Tel says it wants Visible to keep track of foreign social media, and give spies early warnings. Of course, such a tool can also be pointed at domestic bloggers or tweeters. Visible already keeps tabs on Web 2.0 sites for Dell, AT&T and Verizon. For Microsoft, the company monitored the buzz on its Windows 7 rollout. For Spam-maker Hormel, Visible is tracking animal-rights activists' online campaigns against the company.

Emotions have a major impact on essential cognitive processes; neurological evidence indicates they are not a luxury. Emotions play a necessary role not only in human creativity and intelligence, but also in rational human thinking and decision-making. Thus, one fundamental aspect of "listening" is that it isn't just about cold data, it's about observing emotions. The latest scientific findings indicate that emotions play an essential role in decision making, perception, learning, and more—that is, they influence the very mechanisms of rational, cognitive thinking. Thus, the affective domain is needed to balance the cognitive domain throughout the entire OODA cycle. Listening to social networks must go beyond *like* or *dislike*, to assess the full range of frustration, stress, and mood indirectly, through natural interaction and conversation Not only too much, but too little emotion can impair decision making by those we are listening to and by us in making decisions related to our listening posts.

Thus, emotion is not only necessary for creative behavior, but neurological studies indicate that decision-making without emotion can be just as impaired as decision-making with too much emotion.[2]

Affective computing emerged over a decade ago in order to enable intelligent interaction between humans and computers through the development of systems with more human-like computing abilities. According to its founder, Rosalind Picard, M.I.T. professor and author of the 1997 seminal book, *Affective Computing* is computing that relates to, arises from, or deliberately influences emotions. Just as we have become overwhelmed with the cold, hard facts being served up as Big Data, now it seems there's a whole other realm that must be taken into account. *How do you feel about that? Can your database management system handle emotional data? Can your business intelligence system comprehend and, in some cases, feel emotions?*

While developments in the field of affective computing remain largely in the lab, an associated field of endeavor is *opinion mining* and *sentiment analysis.* An important part of our information-gathering endeavors has always been to find out what other people think. With the growing availability and popularity of opinion-rich resources such as online review sites and personal blogs, new opportunities and challenges arise as people now can, and do, actively use information technologies to seek out and understand the opinions of others.

Opinion mining and sentiment analysis, deals with the computational treatment of opinion, sentiment, and subjectivity in text, and is a direct response to the surge of interest in new social networks that deal directly with opinions as first-class objects. In short, we are mining the Web for *feelings*, not *facts.*

Automated sentiment analysis is the process of training a computer to identify sentiment within content through Natural Language Processing (NLP). Various sentiment measurement platforms employ different techniques and statistical methodologies to evaluate sentiment across the Web. Some rely 100% on automated sentiment, some employ humans to analyze sentiment, and some use a hybrid system. According to SenticNet *(www.sentic.net)*, "Existing approaches to opinion mining and sentiment analysis can be grouped into four main categories: keyword spotting, in which text is classified according to the presence of fairly unambiguous affect words; lexical affinity, which assigns arbitrary words a probabilistic affinity for a particular emotion or opinion polarity; statistical methods, which calculate the valence of keywords and word co-occurrence frequencies on the base of a large training corpus; finally sentic computing, which uses affective ontologies and common sense reasoning tools for a concept-level analysis of natural language text."

Companies such as Kia, Best Buy, Paramount Pictures, Cisco Systems, and Intuit are using sentiment analysis to determine how customers, employees, and investors are feeling. Some companies even use software to check the *tone* of e-mail messages and other communications.

Let there be no doubt that the amount of information, especially unstructured information (blogs, tweets, videos et al) is exploding to the point that one of the most urgent topics in business computing is Big Data (data often reaching exabyte scales). Big Data is so impactful that we devote a full chapter, Chapter 13, to it.

For now though, let's consider just one more facet of this discussion of Listening Posts, *Prediction Markets*, which don't necessarily always involve Big Data when they are deployed on just an enterprise-wide basis.

Prediction markets (a.k.a. predictive markets, information markets, decision markets, idea futures, event derivatives, or virtual markets) are speculative markets created for the purpose of making predictions. If Dr. Edwards Deming was right, that all management is effectively prediction, and our traditional methods of prediction are being eclipsed, then prediction markets become a critical innovation management tool. Like financial markets, prediction markets are big information processors, distilling the collective wisdom of their traders (who may be employees of a given company, or the people who participate in public markets) who often may use virtual money to place their bets. The success of any market depends upon the stakes and the pool of traders.

> *"Betting markets don't have any mystical power, but they do summarize conventional wisdom pretty well."* —Professor Paul Krugman.

Prediction markets are not like public opinion polls, in which people vote for who they want to win. The mechanisms for prediction markets are more like betting in the stock market, with knowledgeable people betting on the most likely outcome of an event. Several prediction markets currently exist, such as TradeSports, Iowa Electronic Markets, Hollywood Stock Exchange, HedgeStreet, Yahoo Tech Buzz Game, NewsFutures and Inkling Markets.

Source: tinyurl.com/78gwbw2

When it comes to tapping the collective wisdom of employees in a given company, the author of the book, *The Wisdom of the Crowds*, James Surowiecki, commented that "The success of prediction markets will show that real knowledge is within the organization as a whole and employees have a lot to say about what the future looks like." If all this sounds like idle business fiction or distraction, rather than a means of supporting real-life, serious decision making, consider some examples from Wikipedia …

- Best Buy, Motorola, GE, Qualcomm, Edmunds.com, and Misys Banking Systems are listed as clients of prediction markets software vendor, Consensus Point.
- Hewlett-Packard pioneered applications in sales forecasting and now uses prediction markets in several business units. HP is working toward a commercial launch of a product, BRAIN (Behaviorally Robust Aggregation of Information Networks).
- Corning, Renault, Eli Lilly, Pfizer, Siemens, Masterfoods, Mittal and other global companies are listed as NewsFutures customers.
- Intel is mentioned in *Harvard Business Review* in relation to managing manufacturing capacity.
- Microsoft is piloting prediction markets internally.
- France Telecom's Project Destiny has been in use since mid-2004 with demonstrated success.
- Google has confirmed that it uses a predictive market internally in its official blog.
- The *Wall Street Journal* reported that General Electric uses prediction market software to generate new business ideas.

- *BusinessWeek* lists MGM and Lionsgate Studios as two HSX clients. HSX built and operated a televised virtual stock market, the Interactive Music Exchange for Fuse Network's Fuse TV to be used as the basis of their daily live television broadcast.
- Starwood embraced the use of prediction markets for developing and selecting marketing campaigns. The marketing department started out with some initial ideas and allowed employees to add new ideas or make changes to existing ones. Then subsequently incentives-based prediction markets were leveraged to select the best of the lot.

Persona Management Systems

Okay, so much for the listening part—now you have to act on what you've heard or seen (e.g., YouTube). This brings us back to the bid solicitation from the Department of the Air Force that awarded the $2.76m contract to Ntrepid, a newly formed corporation registered in Los Angeles. This stuff is so super-secret that if you visit Ntrepid's *entire* Web site, this is *all* you will see:

Ntrepid's "persona management software" (PMS) provides licenses for 50 users with 10 personas each, for a total of 500. The personas are replete with background, history, supporting details, and cyber presences that are technically, culturally and geographically consistent. It also provides for secure virtual private networks that randomize the operator's Internet protocol (IP) address, making it impossible to detect that it's a single person orchestrating all the posts. The software also provides static IP address management for each persona, making it appear as though each sockpuppet (fake person) was consistently accessing from the same computer each time. Further, the software uses methods to anonymously establish virtual servers with private hosting firms in specific geographic locations.

This allows a server's *geosite* to be integrated with its social media profiles, effectively gaming geolocation services. The Air Force says that the "place of performance" for the contract would be at MacDill Air Force Base, along with Afghanistan and Iraq.

So here we are, social networks, listening posts and persona management systems. Just as DARPA gave us the Internet, welcome to the new world of the DoD's PMS. Hmm, is it 1984 yet?

A sockpuppet is a false online identity used for purposes of deception within an online community. Through the false identity a community member speaks while pretending to be a different person, like a ventriloquist manipulating a hand puppet. Using sockpuppets, trolls and drones, the military can infiltrate social media to conduct psychological cyber warfare by creating the illusion of consensus. And consensus is a powerful persuader. What has more effect, one person saying that BP wasn't at fault for the Gulf oil spill? Or 500 people or sockpuppets saying it? For many people, the number can make all the difference.

Companies will go to great lengths to master the art of Manufacturing Consent and managing not just back-office processes or enterprise processes or inter-enterprise processes, but now on to managing community-facing "listening-post processes" and "sockpuppet persona processes."

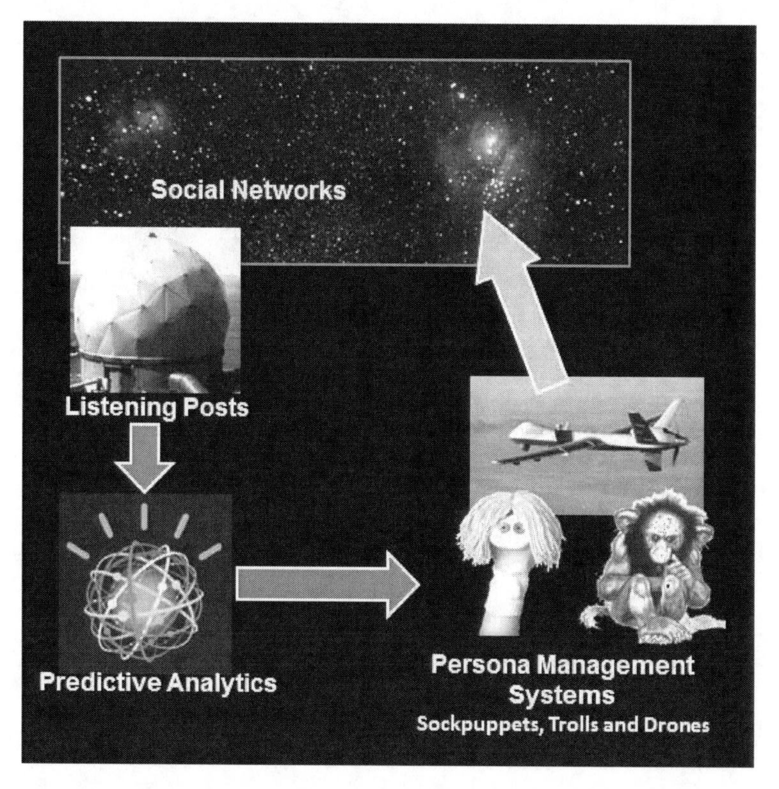

Unleashing sockpuppet processes dives directly into the deep waters of business ethics. But short of a revival of "critical thinking"—a field of study seemingly dropped long ago by our educational systems—get ready to live *in very interesting times* (ancient Chinese curse). Today we seem trapped in a world of sound-bite thinking—no matter the source. We can only hope for the arrival of TMS or Trust Management Systems that provide a counterbalance to the new world of PMS.

Whatever Became of Critical Thinking?

Esteban Colberto made a blog post with some very interesting questions, "My questions to you are:

- Are you a sock puppet?
- How do I know for sure?
- How do you know that I'm not a sock puppet?
- What if we're both sock puppets?
- If we all had sock puppet technology, would we spend all of our time hiding our dirty sock puppets?
- If all we're doing is dirty sock puppet laundry, what's the point of social media?" [3]

So, without an academic discussion, let's turn to Youtube to get a gut feeling for this matter of trust:

"But you didn't say I could have a *real* pony."
"... You didn't ask!" *tinyurl.com/obkqle*

By the way, who really wrote this chapter? In case you wondered, a real human actually did write this chapter, but some of what you might be reading in the future could actually be written by a very linguistically-smart computer. Don't believe us?

As reported by *New York Times* writer Steve Lohr, the news brief was written within 60 seconds of the end of the third quarter of the Wisconsin-U.N.L.V. football game in September 2011. The news brief is the handiwork of Narrative Science that offers proof of the progress of artificial intelligence—the ability of computers to mimic human reasoning. The company's software takes data, like that from sports statistics, company financial reports and housing starts and sales, and turns it into articles, as illustrated in the figure below.

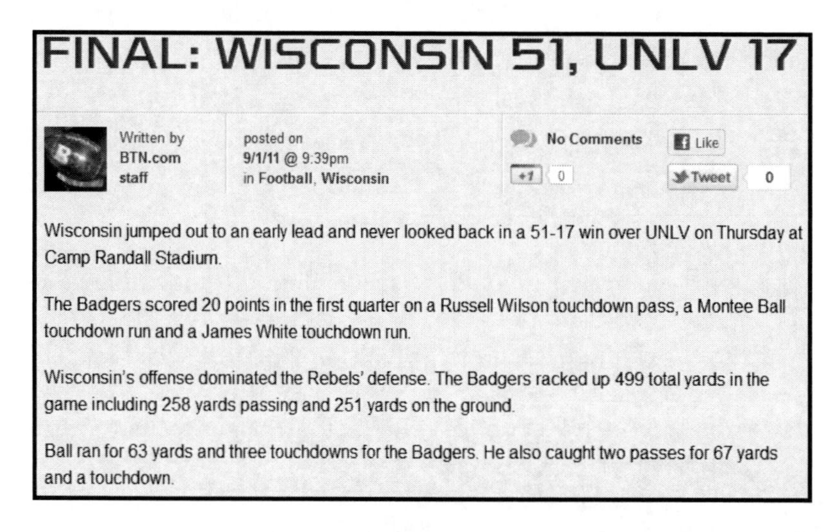

FINAL: WISCONSIN 51, UNLV 17

Written by BTN.com staff

posted on 9/1/11 @ 9:39pm in Football, Wisconsin

No Comments

Like

+1 0

Tweet 0

Wisconsin jumped out to an early lead and never looked back in a 51-17 win over UNLV on Thursday at Camp Randall Stadium.

The Badgers scored 20 points in the first quarter on a Russell Wilson touchdown pass, a Montee Ball touchdown run and a James White touchdown run.

Wisconsin's offense dominated the Rebels' defense. The Badgers racked up 499 total yards in the game including 258 yards passing and 251 yards on the ground.

Ball ran for 63 yards and three touchdowns for the Badgers. He also caught two passes for 67 yards and a touchdown.

Narrative Science is based on more than a decade of research, led by two of the company's founders, Kris Hammond and Larry Birnbaum, co-directors of the Intelligent Information Laboratory at Northwestern University, which holds a stake in the company. Mr. Hammond cited a media maven's prediction that a computer program might win a Pulitzer Prize in journalism in 20 years—and he begged to differ. "In five years," he says, "a computer program will win a Pulitzer Prize—and I'll be damned if it's not our technology."

Perhaps the next breakthrough will be intelligent reading software that we can use to read those computer-generated articles for us!

Takeaway

Today, George Orwell's book, *1984*, written in 1949, is a relevant backgrounder for the 21st century. Chomsky's 1988 book is still in the top rankings at Amazon after 23 years. Read it and extrapolate the information to the new worlds of social media in business.

Who actually wrote this chapter?

References

[1] http://thenextweb.com/apple/2011/10/23/siri-mickey-mouse-and-apples-cult-of-personality/

[2] http://www.jonahlehrer.com/books/how-we-decide

[3] http://forums.colbertnation.com/?page=ThreadView&thread_id=25753

9. Collective Intelligence: The Deliberatorium and Argumentation Systems

With Mark Klein, MIT Center for Collective Intelligence

Social Networks are good for generating and disseminating ideas and knowledge, but there is also a lot of "*noise, disorganized content and polarization.*" And this can lead to a lot of confusion and inefficiency.

To be successful, innovation must be real. In order to be innovative, there must be new ideas or ideas used in new ways. They must be forward thinking, they must be feasible, they must be viable and they must be valuable. The velocity of business dictates rapid innovation, which means get lots of ideas quickly, be able to assess them quickly, and then be able to effectively execute on them quickly.

Information theory is generally considered to have been founded in 1948 by Claude Shannon in his seminal work, *A Mathematical Theory of Communication*. The central paradigm of classical information theory is the engineering problem of the transmission of information over a noisy channel.

Right now, social networks are the ultimate noisy channel. There are two consequences of this noise—efficiency of the communications (how much do I have to listen to in order to grasp what is being communicated) and accuracy (can I still get the right message even if I miss part of the communication). Lastly, even information theory does not address the issues of quality or importance of the information being shared. In order to become more effective and efficient at innovation across the OODA spectrum, these issues need to be addressed.

The "collective intelligence" of members and stakeholders in an organization are not completely harvested by the decision makers even in cases of paramount importance. In fact, in such cases this lack of total harvest can make the difference between a success or a disaster. Social media technologies (e.g. email, web forums, chat rooms, blogs, wikis, and idea forums) have the potential to address this critical gap. Such tools have enabled diverse communities to weigh in on topics they care about at unprecedented scale, in turn leading to remarkably powerful emergent phenomena such as:

- *Idea synergy*: when users share their creations in a common forum, it can enable a synergistic explosion of creativity as people develop new ideas by combining and extending ideas that have been put out by others.
- *The long tail*: social computing systems enable access to a much greater diversity of ideas than they would otherwise: "small voices" (the tail of the frequency distribution) that would otherwise not be heard can now have significant impact.
- *Many eyes*: social computing efforts can produce remarkably high-quality results by virtue of the fact that there are many eyes continuously checking the shared content for errors and correcting them.
- *Wisdom of the crowds*: large groups of (appropriately independent, motivated and informed) contributors can collectively make better judgments than the individuals that make them up, even experts, because their collective judgment cancels out the biases and gaps that individuals are prone to have on their own."

While the social media have enabled an explosion in how much communities can weigh in on topics they care about, they often create more *heat* than *light* when applied to complex, controversial problems:

- *Disorganized content*: Existing social media generally create very *disorganized* content, so it's time-consuming to find what has been said on any topic of interest. This fosters unsystematic coverage, since users are often unable to quickly identify which areas aren't yet well-covered and need more attention.
- *Low signal-to-noise ratio*. Social media content is notorious for producing highly *redundant* content, so important points often got lost in the crowd.
- *Quantity rather than Depth*. Social media systems often elicit many relatively small contributions, rather than a smaller number of more deeply-considered ideas, because collaborative refinement is not inherently supported.
- *Polarization*: Users of social media systems often self-assemble into groups that share the same opinions, so they see only a subset of the issues, ideas, and arguments potentially relevant to a problem. People thus tend to take on more extreme, but not more broadly informed, versions of the opinions they already had.
- *Dysfunctional argumentation*: Existing social media systems do not inherently encourage or enforce well-considered argumentation, so postings are often bias- rather than evidence- or logic-based."

For these reasons we presented some of the lessons learned about open innovation in Chapter 7, particularly:

- Carefully craft the problem statement (it has to be in plain language)
- Experiment in-house first (smaller group)
- Offer incentives
- Set a time limit
- Focus on desired OUTCOMES with a transparent evaluation process

Collecting or harvesting a key fraction of collective intelligence or wisdom of a community from among a veritable glut of information requires enormous effort. The Deliberatorium, developed at the MIT Center for Collective Intelligence, is an advanced software system designed to transcend these limitations and realize the enormous potential social media has for enabling better organizational decision-making. Members of a community are guided to make their contributions in the form of a *deliberation map*, a tree-structured network of posts each representing a single *issue* (question to be answered), *idea* (possible answer for a question), or *argument* (pro or con for an idea or other argument).

A key element of the Deliberatorium is the "live and let live" rule: if one disagrees with an idea or argument, the user should not change that post to undermine it, but should rather create *new* posts that present the strongest ideas and arguments they can muster, so all contributions can compete on an even basis.

Moderators help ensure that the maps are correctly structured. They act as *honest brokers.* Their role is not to evaluate the merits of a post, but simply to ensure that the content is structured in a way that maximizes its value to the community at large.

This process is supported by such software capabilities as *open editing* (any user can check and improve posts), *watchlists* (which automatically notify users of changes to posts they have registered interest in) and *version histories* (to allow users to roll-back and post to a previous version if it has been "damaged" by an edit). The system also provides a powerful set of *attention allocation metrics* to assess how well each part of the deliberation is going, and thereby to help community members focus their efforts where they can do the most good.

Deliberation maps have many advantages over conventional social media. Every unique point appears just once, radically improving the signal-to-noise ratio. All posts appear under the posts they *logically refer to*, so all content on a given question is co-located, making it easy to find what has and has not been said on any topic, fostering more complete coverage, and counteracting polarization by putting all competing ideas and arguments right next to each other.

The Deliberatorium also encourages critical thinking because the users need to present the evidence and logic of the ideas that they generate or favor. And the social community rates the individual's contribution of the ideas. This ultimately leads users to *collaboratively* refine proposed solution ideas.

One user can, for example, propose an idea, a second user can raise an issue concerning how some aspect of that idea can be implemented, and a third can propose possible resolutions for that issue. The value of an argument map can extend far beyond the problem instance it was initially generated for: it represents an entire *design space* of possible solutions that can be readily harvested, refined and re-combined when similar problems arise at other times and places."

The Deliberatorium has been tried out in various business, governmental and educational contexts and found to make a real difference. The first experiment involved translating a Web forum on carbon offsetting into a deliberation map. This illustrated the potential of deliberation maps in harvesting collective intelligence of the community which was, for obvious reasons, qualitatively more useful than the conventional media.

The first large-scale Deliberatorium evaluation was conducted at the University of Naples. 220 students in the information engineering program were asked to weigh in, over a period of three weeks, on what use Italy should make of bio-fuels. All told, the students contributed nearly 2,000 posts, creating a map that was judged by content experts to represent a remarkably comprehensive and well-organized review of the key issues and options around bio-fuel adoption, exploring everything from technology and policy issues to environmental, economic and socio-political impacts.

All this was achieved with the part-time support of just two moderators. We were hard-pressed to imagine an approach that would allow over 200 authors to write what was in effect a substantial book on a complex subject, in a couple weeks, with *no one in charge.*

The first business-centric evaluation was conducted with Intel on the question of how "open computing" (i.e. where users are given greater access to computing tools and data) should be used in the company. Contributions were purely voluntary. A single moderator was able to support the discussion with very little effort. The end result was that Intel received a substantive and well-organized overview of important issues in this space from 73 contributors, including many from outside the company, at essentially *zero cost.*

The MIT Center also conducted evaluations with the U.S. Bureau of Land Management, the University of Zurich, and HMC Inc. etc. All these evaluations have proved that even though Deliberatorium is not the crux of all organizational decision making, it is a better bet than conventional social media and it enables its users to harvest better and complete content on how to solve complex problems. The Deliberatorium also makes it possible to hear significant but small voices and most importantly it makes it easier to find the right stuff, ensuring lack of bias.

The paper, "How to Harvest Collective Wisdom on Complex Problems," focuses on the virtues of question-centric tools: "Another promising social media technology is question-centric tools such as Dell's IdeaStorm, the Obama administration's Open for Questions web site; solution contest sites such as Innocentive, and Google's project10tothe100.com. Such tools are organized around questions: a question is posted and the community is asked to contribute ideas for how to answer that question. Such sites can elicit huge levels of activity."

Interested parties should formulate the following steps, if they want to try out the process of Deliberatorium in their organizations:

1. Embed the tool into a management process so that it's part of the "workflow" and taken seriously (e.g., make it a part of the strategic planning process). Your organization's members are much more likely to try a new approach if it is part of a conversation that *matters.*

2. As an initial test of the tool's effectiveness, conduct a "controlled experiment:"

- Select a specific theme or question that needs to be answered as part of the strategic planning process (e.g., What will be the likely impact of emerging market growth to our business model? How will the competitive environment be changed as a result of a stronger role of states and regulation?)
- Set up two groups that are comparable in terms of their size, diversity and skill sets. In one group, deploy the Deliberatorium to surface perspectives on the strategic planning question. In the other group, run the process in the traditional form. Compare the results in terms of (1) quality/depth of deliberations; (2) quality/depth of output; and (3) level of engagement/satisfaction in the process by the users.

3. If the results for the experimental group are better than those of the control group, further build out the tool and supporting processes, and deploy more broadly at the next opportunity."

Going forward, the Center for Collective Intelligence is working on two areas that require "large-scale argumentation" to support effective large-scale deliberation:

Attention Mediation: The Center is currently investigating how automatically-derived deliberation metrics can enable better attention allocation decisions. This has proved to be a fertile area; deliberation maps offer a unique opportunity to define novel and useful metrics because they capture a significant degree of semantics (post types and links) in addition to the keyword frequency and social network information available with standard social computing tools.

Metrics can be derived that help to identify how mature a deliberation is (deliberations tend to move from issue identification to idea brainstorming to idea evaluation), what the controversial issues are (there are many opposing ideas and arguments, with deep argument chains), whether some elements of the discussion evince a high level of bias (people give high ratings to ideas whose underlying arguments are poorly rated), whether some elements of the discussion have become highly balkanized (there are distinct groups of participants that talk mainly to each other and have opposed positions on a given issue), whether there is excessive groupthink (community attention is focused on a small subset of the options for an issue before the other options have been much explored), and so on.

Moderation: A key challenge with the large-scale argumentation approach is the need for moderators to help ensure that contributors produce a well-structured deliberation map despite the fact that most authors will not be initially familiar with the formalism.

While it is reasonable to expect that super-users with strong moderation skills will emerge from large long-lived deliberation communities, just as they have from other social media communities, this does not help if we have shorter-lived deliberation engagements and, in any case, wider adoption of a large-scale argumentation approach will be facilitated if we can reduce the moderator burden as much as possible.

MIT is exploring, for this purpose, a combination of search tools (to help authors and moderators quickly find where to locate their posts) as well as "wisdom of the crowds" moderation (wherein large numbers of people, performing simplified micro-tasks, substitute for a relatively small cadre of expert moderators).

Takeaway

It has been shown many times that large numbers of people with even modest skills can, under some circumstances, in the aggregate perform judgment tasks better than experts. It has also been shown, by such systems as Amazon's Mechanical Turk,[1] that large number of people are willing to perform such micro-tasks cheaply, or even for free if they believe in the project—*crowdsourcing* in one of its purest forms.

Learn more about the Deliberatorium:

cci.mit.edu/klein/videos/concepts-intro.mov
or tinyurl.com/57tjcj
cci.mit.edu/klein/papers/deliberatorium-intro.pdf
or tinyurl.com/7qb3x4y

References

[1] https://www.mturk.com/mturk/welcome

10. Business Process Innovation in the Intercloud

The business processes that were once tightly confined within a single, vertically integrated company are blown to bits by the Internet and now stretch across multiple companies. These days, over 20 companies make up a typical value chain. Reassembling the business process bits from multiple companies into a coherent infrastructure is the centerpiece of 21st century business architecture. Think value webs, not value chains. The result is a web of any-to-any connections that can drive supply chains, demand chains, and even the business processes that represent the core competencies of an enterprise.

To successfully compete on the battlefields of 21st-century business, companies must reinvent their processes and culture in order to sustain innovative solutions. In the age of *extreme* total global competition, companies must transform their end-to-end business processes, their entire value chains, including their social infrastructures to create ongoing innovation advantages.

In their book, *The New Age of Innovation,* M.S. Krishnan and the late C.K. Prahalad argue that the key to creating value and the future growth of every business depends on accessing a global network of resources to *co-create* unique experiences with customers, one at a time. To achieve this, CEOs, executives, and managers at every level must transform their business processes, technical systems, and supply chain management systems. Prahalad and Krishnan explain how to accomplish this shift—one where IT and the management *architecture* form the corporation's fundamental foundation. As we do in this book, they argue that this architecture must provide a foundation for:

- Redesigning systems to co-create value with customers and connect all parts of a firm to this process
- Measuring individual behavior through smart analytics
- Ceaselessly improving the flexibility and efficiency in all customer-facing and back-end processes
- Treating all involved individuals—customers, employees, investors, suppliers—as unique
- Working across cultures and time-zones in a seamless global network
- Building teams that are capable of providing high-quality, low-cost solutions rapidly

Companies must manage not just a single value chain, but multiple, individual value chains—simultaneously. The business processes that drive each value chain must be light weight and have the right level of granularity. That's where SOA comes in. With a SOA foundation, business process components can be bundled, unbundled and rebundled as "End-to-End Situational

Business Processes" in response to new threats or opportunities. They can allow a company to participate in multiple marketplaces or reach out directly to individual customers with personalized offerings. Moreover, they take the risk out of business innovation as innovation initiatives that involve multiple business partners can be scaled up or down in a heartbeat.

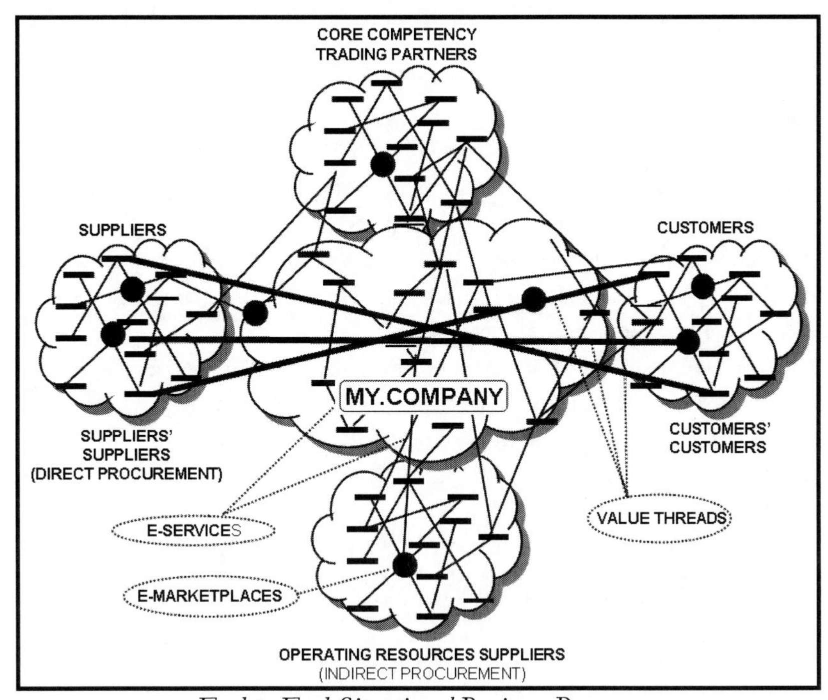

End-to-End Situational Business Processes
Supporting Multiple, Simultaneous Value Chains
(Source: *Dot.Cloud: The 21st Century Business Platform)*

Because failure is just as important to business innovation as success, the Cloud provides the platform to fail early, shut down a failed initiative and move on again with lessons learned.

Now the intriguing part. With multiple players involved in any given value chain, the need is apparent for not just one cloud, but multiple interoperating Clouds, or the Intercloud, if you will.

In the era of mass customization, personalization and multiple innovation initiatives, a unique value chain will no doubt be needed for each initiative. It's even possible that just a single customer might be involved in a given initiative and the resulting value chain (let's say VC-1). Others may serve multiple customers over long periods of time (let's say VC-2). Companies will need to manage both kinds of value chains and their underlying end-to-end processes for each *situation* (hence SBPs). Variations in sourcing, customer service support, products, and market channels will be unique to each value chain.

> This dynamic 21st century demand-supply value chain
> is the stuff of business innovation in the Cloud.

Just consider UPS. The company provides a world-class parcel delivery service, as we all know. But when the company got into the computer repair business, delivery was only a small part of the new value chain UPS would have to support. When a Toshiba laptop breaks down due to a hardware failure, UPS certified technicians in Kentucky do the repairs, requiring a whole new set of value chain resources and activities that must be managed.

The figure below depicts *hypothetical*, end-to-end, Intercloud, situational business processes used to launch a killer new product. It's a hybrid Cloud mashup that's ready for a consumer swarm driven by Facebook and other social networks.

Thought the *iPhone* was hot? Then the *Android phone*? Welcome to the sg-phone, the sun glasses phone that will obsolete all others. The heads-up display, ear pods and eye-wink cursor control, backed by voice commands, does it all. Conceived by a scientist gone mad, Warren Buffet's Berkshire Hathaway immediately invested $20 billion on the condition that the product be brought to market in an instant.

How'd you like to be the CIO that could make such possible?

Well, Claude Skydancer had a plan.

On the design side, he initially turned to Yet2.com to source various designs for electronic components. On the design and production side, he would plug in the legendary original design manufacturer (ODM), Flextronics in Singapore, via the e2open Community Cloud, which would, in turn, tap its suppliers using Avaro's Supplier Information Management as a Service (the same company GE uses) to source and manage any number of Tier-n component suppliers. Flextronics and its suppliers would also tap e2open's business network (Community Cloud) for B2B interactions, as well as Workday's human capital management (HCM) service for its 200,000 employees scattered over 30 countries.

Because the trading partners operate on their own clocks, My.Company's Business Process Management System (BPMS), hosted in the Cloud, serves as the "choreographer" of the entire value delivery system. Further, the BPMS provides real-time business activity monitoring (BAM) that's essential to continually optimize the overall operations of the overall value delivery system.

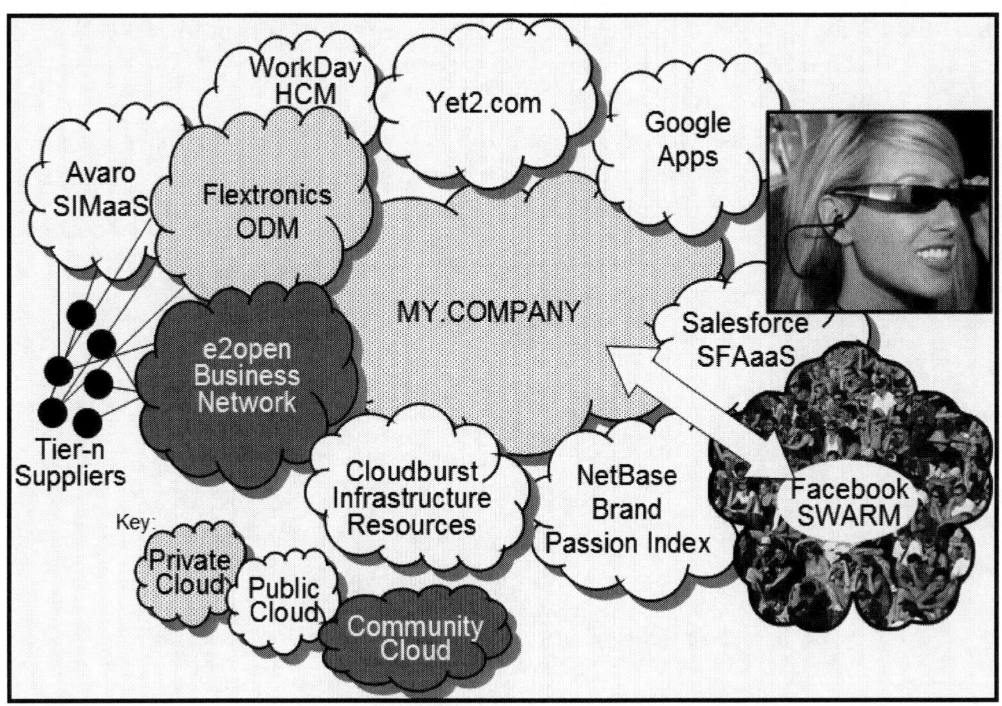

End-to-End Business Processes in the Intercloud
(Source: *Enterprise Cloud Computing)*

A community or privately hosted BPMS must incorporate both *orchestration* and *choreography*. With orchestration, there is a centralized control mechanism that directs activities, each of which is an interaction between services. While orchestration is how one executes a composition of services, it isn't the method by which one composes services. Choreography has no centralized control. Control is instead shared between multi-company participants in a peer-to-peer model. Choreography is at a higher level and looks at a process as message-passing between peers, without delving into the processes internal to any of the participants.

Because the typical value chain is composed of multiple companies, companies that run under their own internal control and against their own internal clocks, the notions of orchestration and choreography come front and center. While the Cloud literature is full of writings on mashup applications, the real challenge for the enterprise is orchestrating and choreographing "services" into end-to-end business processes that deliver value to the customer. Once again, that's where business process management delivered via the Cloud comes in.

Meanwhile NetBase's social network analytic services are used to listen to customer communities and maintain the brand's "passion index." Google maps were mashed up to bring retailer locations into the picture. Public infrastructure services were put on standby for "cloud bursting" when the product launch became overwhelmed by a Facebook swarm and other social computing media.

The Intercloud won't just "happen" one Monday morning with a master switch in the sky being swirched on. It will be a process of step-by-step transformations requiring standards as various industries move from the current "As-Is" world of IT to the "To-Be" state.

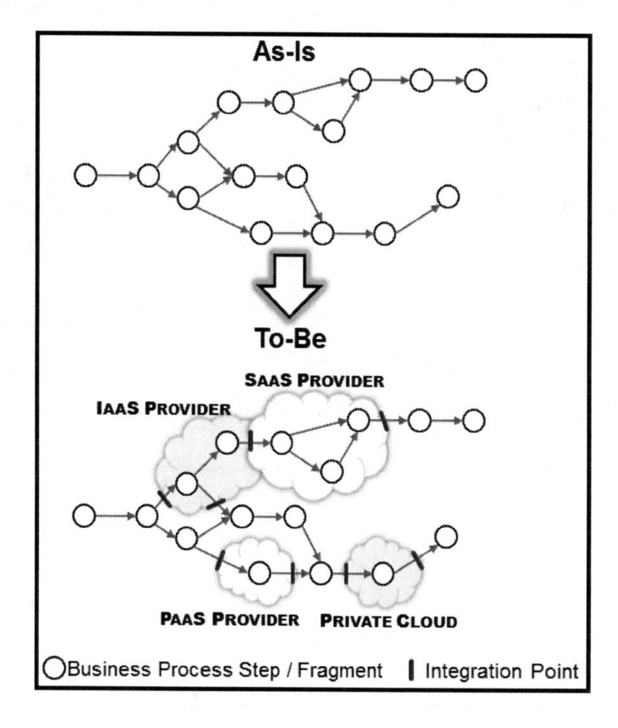

Sooner than many expect, companies will shift from 100% on-premises, 0% Cloud-enabled computing resources to a fully integrated, optimal mix of on-premises and Cloud environments. This is, however, a journey, not something that will magically happen next Monday morning because the needed interoperability standards must be developed and adopted—no easy task.

Just as standards were the key to the seamless interoperation of the Web and the Internet itself, standards will be key to the seamless interoperation of the Intercloud. In the meantime we'll see the rise of the Cloud Broker that will function much like today's Systems Integrators for IT systems, but with their expertise centered in the Cloud. We'll explore the role of Cloud Brokers in more detail in Chapter 16.

It's a whole new business world in the Cloud!

Takeaway

Just as the Internet is a network of interoperating networks, the Intercloud is a network of interoperating clouds. Just as Amazon disrupted the retail industry by early adoption and business mastery of the Web, smart businesses, regardless of their industries, are already hard at work seeking to disrupt their industries by leveraging the Cloud. They are already engaged in the business dog-fight in the Cloud and mastering their OODA loops to gain the agility they need.

Caveat competitor!

11. Business Innovation: Culture and the Process

"Don't tell me you want a bridge, show me the canyon you want to cross and tell me why you want to cross it." —Anonymous Designer.

"Despite the roar of voices equating strategy with ambition, leadership, vision, or planning, strategy is none of these. Rather, it is coherent action backed by an argument. And the core of the strategist's work is always the same: discover the crucial factors in a situation and design a way to coordinate and focus actions to deal with them."
—Prof. Richard Rumelt, UCLA Anderson School of Management.

People talking about innovation remind us of the blind men and the elephant. Each has his own perspective about what the elephant is or is like. Each right and each wrong. When the experts talk about innovation, they are often not very helpful. Sometimes they are describing innovation as they look through a telescope, sometimes as they look through a microscope and sometimes they are just looking (though often with rose colored glasses). Also, terms are used so loosely. Describing something as an innovation is like describing either an astronaut with a PhD in astrobiology and an MD with a specialty in cardiovascular surgery as a mammal. How comfortable are you if I say a mammal will be performing your surgery today. The amounts of information conveyed by the two descriptions are radically different.

A Quick Recap

Let's do a quick recap of some of the dimensions of business innovation:

- There are two elements to an innovation organization – a culture of innovation and a process of innovation
- There are three degrees of innovation – Sustaining, Breakthrough and Disruptive. (Chapter4)
- There are three domains of innovation – the Ecosystem, the Enterprise (capability) and the Customer (outcomes).
- There are three forms of innovation – Business (organizational operating) model, Business process (how things happen), and Product / Service
- And there are an unlimited number of types of innovation: Process, Product, Service, Technology, Platform, Application (not software but how something is used), Design, Integration /

combination, Value, Marketing, Experiential, etc.

There are two elements of success in being an innovative organization: Culture and Process. Playing off of Peter Drucker's quote, "Culture eats strategy for breakfast," let's turn our focus first on the Culture of Innovation.

The Culture of Innovation

The bedrock of business innovation is to have a culture of innovation in place within the organization—the sum total of values, norms, assumptions, beliefs and ways of working built up by a group of people and transmitted from one generation to another within the organization. Innovation is required in everything that is done by the company in the 21st century, not just in marketing or in new product development. Innovation in an organization goes beyond simply responding to change—it creates change in the environment that other organizations must respond to, and therefore can become a sustainable competitive advantage. A key factor in boosting innovation is the right environment to nurture, organize and use the creative potential, knowledge of customers, competitors, and processes that employees have (see the discussion of *Employees First* in Chapter 14).

As we keep reminding throughout this book, R&D is not innovation. Booz & Company issued its annual report on the Global Innovation 1000 with the following highlighted, "There is no statistically significant relationship between financial performance and innovation spending, in terms of either total R&D dollars or R&D as a percentage of revenues. Spending more on R&D won't drive results. The most crucial factors are strategic alignment and a culture that supports innovation." Paraphrasing Drucker's quote above, you could say that Innovation Culture eats R&D for lunch.

Processes can be measured, but how do you measure an environment? Some of the ideas of how to measure an innovative culture include innovation propensity, organizational constituency, organizational learning, creativity, empowerment, market orientation, value orientation, and implementation context.

Trust is the first principle of an innovation culture. If you cannot or are unwilling to create trust, then don't make the effort to go further. And while trust is a two-way street, if your organization is based on command and control, then it is incumbent upon management to take the first step and to continue to take steps to mitigate distrust. Later in the book we talk about the end of management, but trust is a precursor to that (even though we argue that it is inevitable).

It's also trusting not just in each other, but in the outcome. The economic model we have today abhors certainty. Even the strongest, most innovative company will fail and fail again. Every innovation contains the seeds of change that spells the end for certain types of jobs and certain positions. If that is not understood and accepted, then the willingness to innovate is dissipated.

The second, organizational principle of an innovation culture is *networking*. We talk about that extensively in this book. While there will continue to be a role for command and control—where priority is to muster resources, delegate and divide work—organizations simply cannot respond fast enough to the rate of change in today's economy. When we talk about processes, OODA loops, and scenario planning—we talk about these concepts in the context of open, transparent operating networks. Networks emerge, evolve, adapt, adopt, and always seem to have an option to "get the work done." Networks are good at proliferating dozens of experiments, testing them in ad-hoc fashion and distributing the best of them in cascades across the organization (see Chapter 12, Innovation is a Team Sport and Chapter 18, The Fractal Company).

The third principle of an innovation culture is that it embeds the notion of a *"gift economy"* where valuable goods and services are regularly given without any explicit agreement for immediate or future rewards. Much like the Native Americans (potlatch), traditional scientific research and the open source community, innovation cultures thrive on networks, and networks coalesce around the exchange of something of value in the eyes of its members. The value must in part be intrinsic: the members of the network find value from inside themselves that compels them to belong. If the value equation relies too much on extrinsic motivation, like paying people for their involvement, its success is more problematic.

The fourth principle of an innovation culture is *transparency*. The network needs a way for evaluating ideas and the idea makers—quality, popularity, affirmation and reinforcement. This creates an effective feedback loop, reinforces the "fairness" of the gift culture and motivates others to contribute. This transparency also means anyone can run with the ideas; *damn the suggestion box.*

This in turn requires the fifth principle, *simple rules of engagement*—agreed to behaviors and universal expectations (see social network discussions). This also includes the fact that participants in the innovation culture cannot abdicate their responsibility to share some of the risk of innovation. Likewise the organization must provide policies, practices and resources that acknowledge this risk and provide space for experimentation, as well as shoulder the rest of the risk—transparently.

The sixth key principle of an innovation culture is that it has a core *innovation process* that it follows.

The Innovation Process

One of the key components within an innovation program is a process around how it should function. By thinking through and designing an innovation process, we must be able to "test" the process theoretically via "what if" scenarios. By conducting an iterative approach to process design, we can identify bottlenecks, breakdowns, system requirements, opportunities for automation and standardization, as well as resources required to handle expected volumes.

Refer back to the Innovation Architecture discussed in Chapter 6. An architecture is the use of abstractions and models to simplify and communicate complex structures and processes to improve understanding and forecasting the behavior of the system. There we presented a way to think about constructing an innovation process.

Such innovation processes also force us to think of ways that innovation opportunities would, and should, come to core innovation teams. There is not one single widely accepted process for innovation. Various companies have evolved their innovation processes over time. Some have many detailed steps. Others are high-level guidelines. But regardless of the chosen process, it must be repeatable and disciplined.

There are many innovation processes out there, but all encompass the following steps described in Chapter 2 and repeated here for quick reference:

1. *Understand and scope* opportunities to identify your most critical assumptions, usually around what problem you are trying to solve.
2. *Capture ideas* to develop the minimum product and service concepts to deliver value (desired, sought, identified, anticipated outcome) that will allow you to learn about those assumptions and deliver the needed outcomes.
3. *Evaluate and select* by testing ideas in the ecosystem of partners, suppliers, and customers to find out if what you believe is shared.
4. *Develop and experiment* as quickly and efficiently as possible until you discover the right answers to delivering the outcomes.
5. *Implement* quickly and then scale iteratively, openly, and collaboratively to the right product and market fit.
6. *Champion* innovative products and services in order to minimize the time the ecosystem has to climb the learning curve to understand the benefit. This isn't about *sales;* it includes the internal organization, suppliers, partners, and customers—the entire ecosystem.

Below are three high-level innovation processes developed by PWC and Gartner, followed by a more detailed process approach.

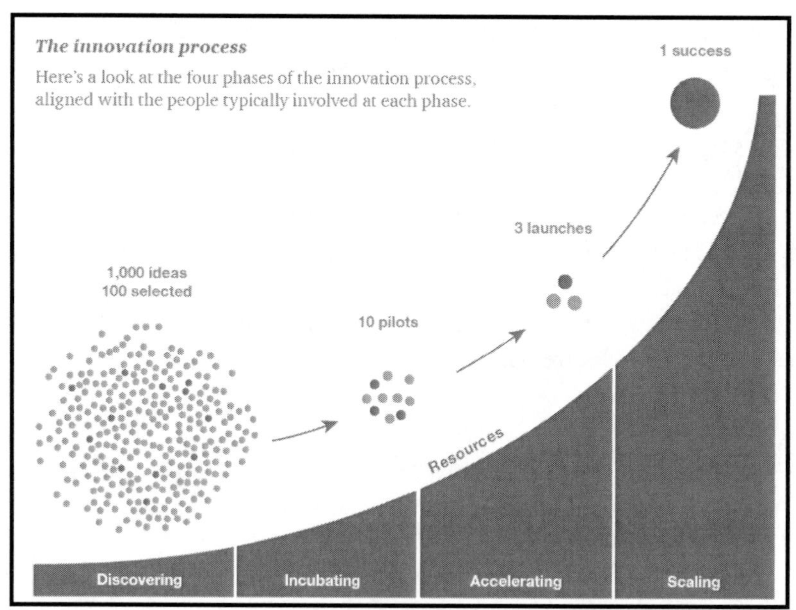

PwC

Phase/Outcomes	OI Techniques	Examples
Phase: Generate Ideas		
• Concepts & ideas	• Source to communities	• Cisco crowdsourcing
• Components	• Acquire from idea markets	• yet2.com
• Patterns	• Web-based pattern seeking	• Social network analysis
Phase: Evaluate & Select Ideas		
• Prototyping	• Acquire prototypes; scale up	• P&G entrepreneur network
• Comment, extend	• Participants review, rank	• Lego User Group
• Rank, vote, select	• Use prediction markets	• Best Buy Prediction Markets
Phase: Develop & Implement		
• Co-develop	• Partner to develop	• Co-research on Alzheimers
• Outsource	• Source full development	• InnoCentive challenges
	• Source key components	• TopCoder ESPN Challenge
• License IP to others	• License solution/components	• Xerox PARC

Gartner

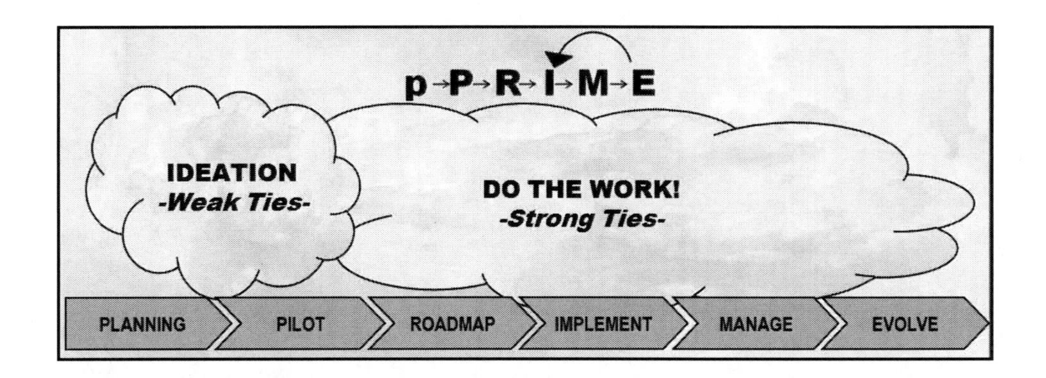

Regardless of the specific innovation process chosen, there are two parts to successful execution—*doing the right thing*, then *doing it right*. The *doing the right thing* part would be an innovation architectural model, the *doing it right* would be the OODA loop model (always adjusting). Below we present an adaptation of the OODA loops, presented earlier in the book, that focuses in on an innovation process lifecycle.

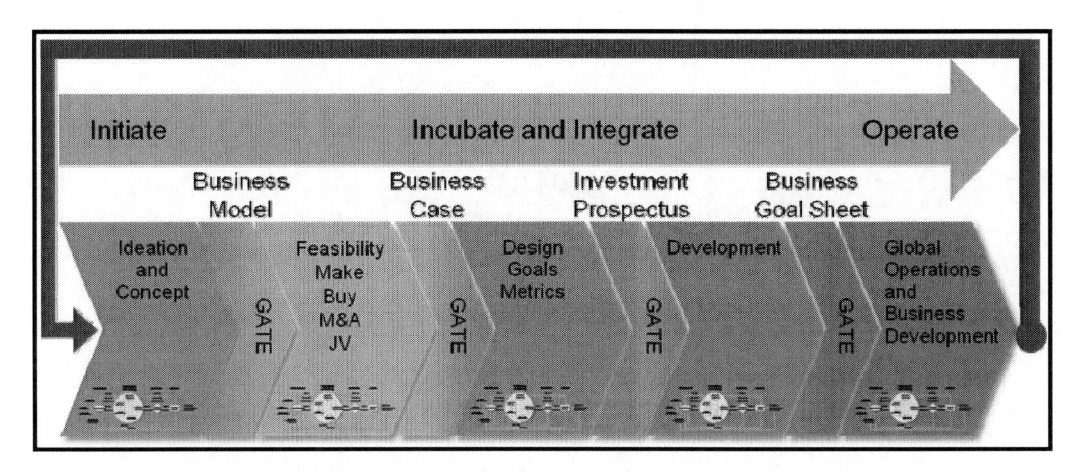

The key to the five steps at the bottom of the model is that each step embeds or nests its own OODA loops. Each step, including the Idea and Concept step, requires adjusting as new information becomes available. And as it is in our discussion of fractals (Chapter 18), each step can influence the overall process. The OODA loop diagram for the five main steps is repeated here from Chapter 5 for quick reference.

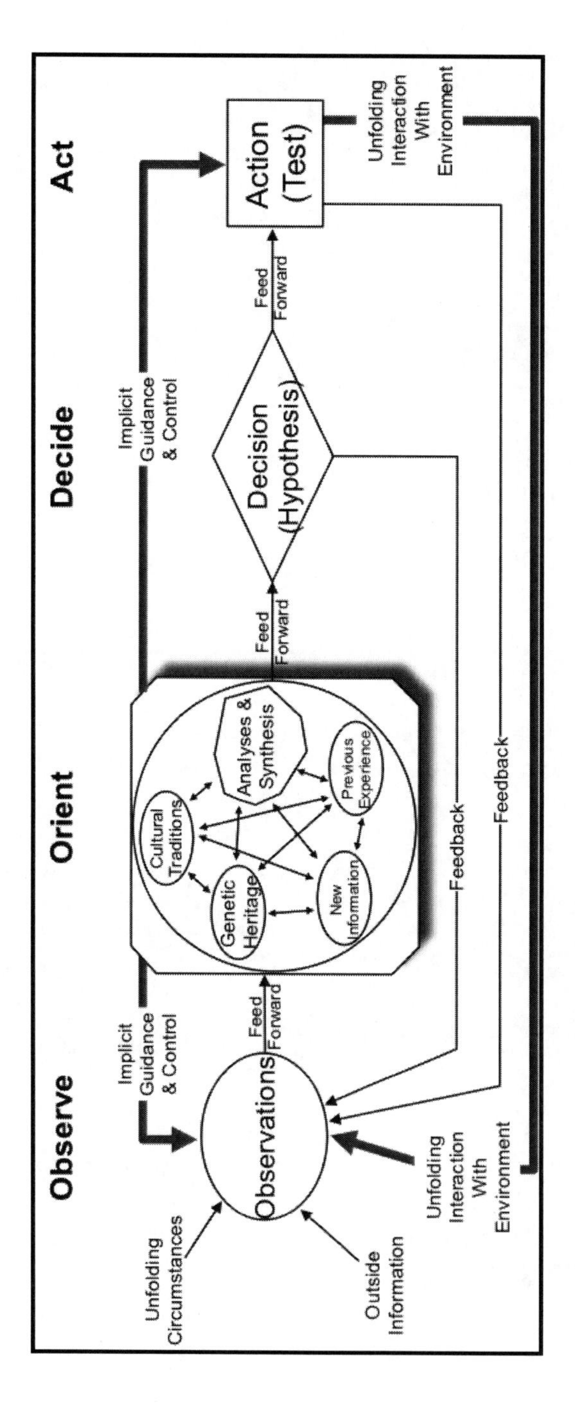
Observe
Orient
Decide
Act
Unfolding Circumstances
Outside Information
Observations
Implicit Guidance & Control
Feed Forward
Cultural Traditions
Genetic Heritage
New Information
Previous Experience
Analyses & Synthesis
Feed Forward
Implicit Guidance & Control
Decision (Hypothesis)
Feed Forward
Action (Test)
Unfolding Interaction With Environment
Feedback
Feedback
Unfolding Interaction With Environment

A business innovation initiative starts with building processes, systems, education programs, and funding that can be integrated into day-to-day operations as well as the culture of a company.

An innovation program can foster stronger competitive advantage, contribute to a company's recognition as an industry leader, and increase customer satisfaction, while also helping business partners be more competitive in their fields. Let's take a quick look at the building blocks and types of plans needed to launch an innovation program.

Building Blocks for Sustainable and Systemic Innovation

Being innovative means being competitive and bringing value to both a company's trading partners and its customers. On one hand, a company can continue to reduce its costs and remain competitive, but today's customers want more than simply reduced costs. On the other hand, a company's business partners want to team with a company that can enable them, and the entire value chain, to be competitive. Cutting costs is just one small part of the equation. Adding unique value is the critical variable in the competitive advantage equation.

There are several building blocks that are required to make business innovation happen, not just once in a while, but over and over again. These foundational building blocks include:

- Create a systemic culture of innovation starting at the C-level through to front-line workers
- Understand how creativity and innovation work
- Establish an Innovation Group within and a commonly understood architecture
- Develop an innovation process and infrastructure, especially tools and capabilities that foster collaboration and a culture of innovation as described above
- Redesign business processes to co-create value with customers and trading partners
- Connect the entire value delivery system to end-to-end business processes
- Work across countries and time-zones in a living ecosystem
- Create multi-company virtual enterprise networks that are banded and disbanded with the ebb and flow of business
- Deploy high-performance teams whose communication protocols are based on information bursts instead of commands (Bioteams, see Chapter 14)
- Understand the requirements that are necessary to make the vision a reality, including reaching outside the company to gather and assess ideas, and then "do the work"

> The critical factor is not
> continuous executive-level support and sponsorship;
> it's total executive immersion and commitment to new business models.

Business innovation goes beyond just funding an initiative and on to the committed belief in systemic innovation throughout the organization. Just consider Google, Apple, Dell and Amazon. Their CEOs didn't delegate "innovation." Larry Page, Steve Jobs, Michael Dell and Jeff Bezos are, or were, in effect, their companies' CIOs (Chief Innovation Officers)!

With executive commitment, large companies can empower thousands of employees to contribute to business innovation, everywhere, everyday.

The executive team needs to understand that failures are expected within an innovation program. They cannot go into a given initiative believing that every endeavor will result in great success.

But the good news is that, thanks to cloud computing, a *sandbox* can be built in no time to test out potentially winning innovations. Google is famous for this approach by releasing beta versions of new offerings and letting users determine their fates.

Hmm? What ever happened to Google Wave? Oops, as Google's site explains, "Google Wave is no longer being developed as a standalone product." Stay tuned.

That's okay, for in a culture of innovation, where a company is doing something never done before, what can be expected? A hole in one on every tee off?

However, the Cloud makes experimenting in a sandbox a no-brainer, with negligible costs and resources—and that's indeed a really big "*Wow*" about cloud computing.

Plan of Intent (Observe and Orient)

Because not every innovative idea, proof of concept, or solution will be successful, it's best to group innovation programs into two categories, Plan of Intent and Plan of Record. Earlier we referred to a Plan of Investigation that is essentially research into "what else do we need to research, what else do we need to watch, what are we missing, what do we don't know we don't know," rather than a specific course of action leading to innovation outcomes as are The Plan of Intent and the Plan of Record.

The Plan of Investigation is a feedback loop into Foresight and Insight. The Plan of Intent involves innovation areas where a company will investigate, research, engineer and document various forms of innovations it expects it will need. These types of projects are more proactive in nature and allow the company to look at various trends, social media and industry observations, then assimilate that information to make recommendations as to where the company should be focusing attention on developing future capabilities.

Scenario planning is key to developing a Plan of Intent. Giving due credit to the military, scenario planning is an adaptation of methods used by military intelligence. For those readers unfamiliar with scenario planning, let's tap Wikipedia for a quick run down.

"Most authors attribute the introduction of scenario planning to Herman Kahn through his work for the US Military in the 1950s at the RAND corporation where he developed a technique of describing the future in stories as if written by people in the future. Scenario planning may involve aspects of systems thinking, specifically the recognition that many factors may combine in complex ways to create some surprising futures (due to non-linear feedback loops). The method also allows the inclusion of factors that are difficult to formalize, such as novel insights about the future, deep shifts in values, unprecedented regulations or inventions. Systems thinking used in conjunction with scenario planning leads to plausible scenario story lines because the causal relationship between factors can be demonstrated. In these cases when scenario planning is integrated with a systems thinking approach to scenario development, it is sometimes referred to as structural dynamics."

"The scenarios usually include plausible, but unexpectedly important situations and problems that exist in some small form in the present day. Any particular scenario is unlikely. However, future studies analysts select scenario features so they are both possible and uncomfortable. Scenario planning helps policy-makers to anticipate hidden weaknesses and inflexibilities in organizations and methods. Here's how military scenario planning or scenario thinking is done:

1. *Decide on the key question to be answered by the analysis.* By doing this, it is possible to assess whether scenario planning is preferred over the other methods. If the question is based on small changes or number of elements, other more formalized methods may be more useful.

2. *Set the time and scope of the analysis.* Take into consideration how quickly changes have happened in the past, and try to assess to what degree it is possible to predict common trends in demographics, product life cycles. A usual timeframe can be five to 10 years.

3. *Identify major stakeholders.* Decide who will be affected by the possible outcomes. Identify their current interests, whether and why these interests have changed over time in the past.

4. *Map basic trends and driving forces.* This includes industry, economic, political, technological, legal, and societal trends. Assess to what degree these trends will affect your research question. Describe each trend, how and why it will affect the organization. In this step of the process, brainstorming is commonly used, where all trends that can be thought of are presented before they are assessed, to capture possible group thinking and tunnel vision.

5. *Find key uncertainties.* Map the driving forces on two axes, assessing each force on an uncertain/relatively predictable and important / unimportant scale. All driving forces that are considered unimportant are discarded. Important driving forces that are relatively predictable such as demographics can be included in any scenario, so the scenarios should not be based on these. This leaves you with a number of important and unpredictable driving forces. At this point, it is also useful to assess whether any linkages between driving forces exist, and rule out any "impossible" scenarios such as full employment and zero inflation.

6. *Check for the possibility to group the linked forces* and if possible, reduce the forces to the two most important, to allow the scenarios to be presented in a neat xy-diagram.

7. Identify the extremes of the possible outcomes of the two driving forces and check the dimensions for consistency and plausibility. Three key points should be assessed:

- *Time frame:* Are the trends compatible within the time frame in question?
- *Internal consistency:* Do the forces describe uncertainties that can construct probable scenarios?
- *The stakeholders:* Are any stakeholders currently in disequilibrium compared to their preferred situation, and will this evolve the scenario? Is it possible to create probable scenarios when considering the stakeholders?

8. Define the scenarios, plotting them on a grid if possible. Usually, two to four scenarios are constructed. The current situation does not need to be in the middle of the diagram (inflation may already be low), and possible scenarios may keep one (or more) of the forces relatively constant, especially if using three or more driving forces. One approach can be to create all positive elements into one scenario and all negative elements (relative to the current situation) in another scenario, then refining these. In the end, try to avoid pure best-case and worst-case scenarios.

9. Write out the scenarios. Narrate what has happened and what the reasons can be for the proposed situation. Try to include good reasons why the changes have occurred as this helps the further analysis. Finally, give each scenario a descriptive (and catchy) name to ease later reference.

10. Assess the scenarios. Are they relevant for the goal? Are they internally consistent? Are they archetypical? Do they represent relatively stable outcome situations?

11. Identify research needs. Based on the scenarios, assess where more information is needed. Where needed, obtain more information on the motivations of stakeholders, possible innovations that may occur in the industry and so on.

12. Develop quantitative methods. If possible, develop models to help quantify consequences of the various scenarios, such as growth rate, cash flow etc. This step does of course require a significant amount of work compared to the others, and may be left out in back-of-the-envelope-analyses.

13. Converge toward decision scenarios. Retrace the steps above in an iterative process until you reach scenarios which address the fundamental issues facing the organization. Try to assess upsides and downsides of the possible scenarios.

"Scenarios planning starts by dividing our knowledge into two broad domains: *(1) things we believe we know something about* and (2) *elements we consider uncertain or unknowable.* The first component – trends – casts the past forward, recognizing that our world possesses considerable momentum and continuity. For example, we can safely make assumptions about demographic shifts and, perhaps, substitution effects for certain new technologies. The second component – true uncertainties – involve indeterminables such as future interest rates, outcomes of political elections, rates of innovation, fads and fashions in markets, and so on. The art of scenario planning lies in blending the known and the unknown into a limited number of internally consistent views of the future that span a very wide range of possibilities."

Scenario planning can be augmented with other tools and methods. For example, when exploring things *we believe we know something about,* companies such as GE turn to "prediction markets." On the other hand, seeking *the uncertain or the unknowable* in scenario planning can be augmented or informed by the Delphi method, yet another innovation brought to us by the military during the Cold War to forecast the impact of technology on warfare.

Delphi is based on the principle that forecasts or decisions from a structured *panel of experts* are more accurate than those from unstructured groups. The Delphi method is a judgmental forecasting procedure in the form of an anonymous, written, multi-stage survey process, where feedback of group opinion is provided after each round.

The Delphi method has instrumental value in providing different alternative futures and the argumentation of scenarios. Because scenario planning is "information hungry," Delphi research can deliver valuable input for the process.

There are various types of information output of Delphi that can be used as input for scenario planning. Researchers can, for example, identify relevant events or developments and, based on expert opinion, assign probabilities to them.

Moreover, expert comments and arguments provide deeper insights into relationships of factors that can, in turn, be integrated into scenarios afterwards. Also, Delphi helps to identify extreme opinions and dissent among the experts.

Such controversial topics are particularly suited for extreme scenarios or wildcards. Organizations can use Delphi in order to make their scenarios more profound and to create confidence in scenario planning. Further benefits lie in the simplification of the scenario writing process and the deep understanding of the interrelations between the forecast items and social factors.

As a form of "collective intelligence" the Delphi method is a natural for social networks in the Cloud. That's precisely where Professor William E. Halal and his associates took the concept and developed the TechCast forecasting system. All organizations need technology forecasts for their strategic planning because technology drives the creative destruction of markets, introduces disruptive products and services, and alters the way organizations work.

Managers often try to develop their own forecasts, but the time and cost are considerable and the results mediocre. The website (techcast.org) now approaches a million hits per year and taps the collective intelligence of experts in over 70 technology fields.

Organizations such as the Rockefeller Foundation offer publicly available *macro-level scenarios* that can be incorporated into individual company scenario planning efforts and initiatives.

THE ROCKEFELLER FOUNDATION

tinyurl.com/2wx4rs5

Scenario planning should be the backdrop of the "Observe" phase of any business innovation initiative for it stretches and refocuses thinking—and *creative destruction* now operates at Internet time. The scenario portfolio can provide an organization with multiple frames of reference to embrace uncertainty and dynamically adjust to *unexpected change*—the kind of exponential change that's no longer to be *unexpected*.

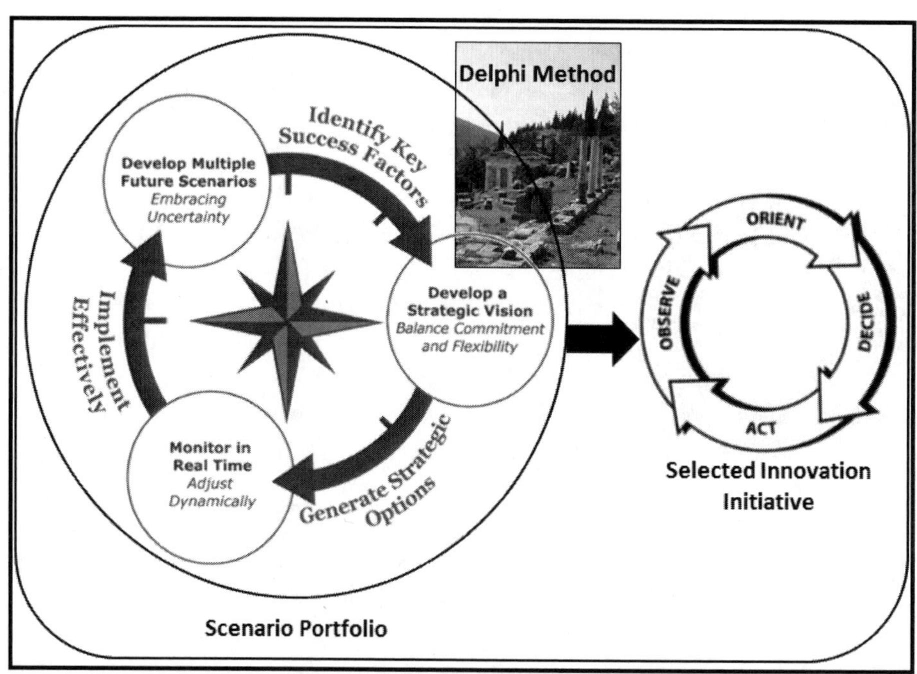

Scenario Portfolio ---> Selected Innovation Initiative
Source: www.sqaki.com

Plan of Record (Decide and Act)

The Plan of Record allows a company to identify specifically where it will invest and commit to deliver results. These innovation projects tend to be more reactionary in nature and have a lower risk associated with them because the company can clearly see customer needs and demands for those categories of innovative solutions.

The objective as an industry innovator is to help ensure that both customers and suppliers are successful and able to grow and thrive. To achieve that, a company must recognize the need to work with a high degree of collaboration at every level, whether in solution design, operational integration, service-level management or change management processes. By working together with its customers and suppliers, a company can expect to identify innovative solutions that can help it meet its current and future business objectives.

To fulfill the role as a key influencer of innovation with our customers and suppliers, a company should *jointly* develop Innovation Plans with its customers and suppliers. An Innovation Plan includes the six components shown below.

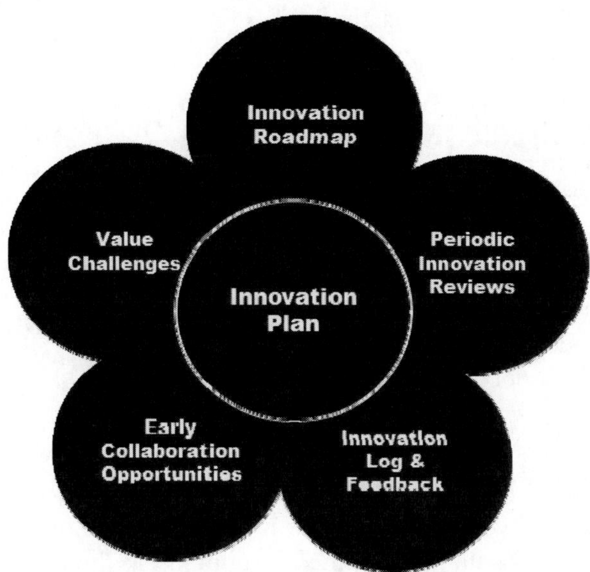

Innovation Plan

If there is one truism to management it is this sequence – you get what you reward, you reward what you manage, you manage what you measure. For innovation to be a systemic, repeatable, predictable process, then some type of management system needs to be in place. Also, nothing happens unless it is on someone's goal sheet, so all organizations should have a group dedicated to enabling, facilitating and accelerating innovation via education, training, mentoring and providing support systems for the innovation process and planning.

Business innovation is only achieved if it brings value to a company's customers and trading partners. Having a comprehensive and well-maintained plan facilitates the implementation of innovative ideas, products and services that bring real value. The Innovation Plan is a resource and a guide that documents the high-level processes the multi-company team uses to identify viable innovation opportunities. It is used as a governance resource as well as a resource for new members of the team to understand the current status of innovation initiatives.

When the first Innovation Plan is collaboratively produced, it will consist of an Innovation Roadmap of items that are set to be implemented as part of the initial initiative as well as a view of possible innovations that could be implemented in the future. Typically, the Innovation Plan is updated at specified intervals, but may be updated more frequently as business conditions warrant.

The Innovation Plan details the processes that are used to regularly identify ideas to review, logs and tracks the innovative ideas that are presented and implemented, and includes a feedback loop to help refine and tune the plan. For emerging or anticipated issues for which there is no planned solution in the Innovation Plan of Record or Plan of Intent, a Value Challenge can be submitted to help solve the problems proactively.

Innovation Roadmap

- To assist customers and business partners in solving their business challenges and introduce innovative ideas into their environments, a company should produce and maintain an Innovation Roadmap. The Innovation Roadmap is a key component of the overall Innovation Plan. It is a timeline consisting of both planned and tentative innovations that will or could be introduced into their environment.
- Items listed on the Innovation Roadmap can include existing solutions, products and services that are on the Innovation Plan of Record and Plan of Intent, as well as relatively undefined solutions for emerging or anticipated issues.

Periodic Innovation Reviews

- To help ensure a sustained focus on innovation and that customer innovation needs are being met, the innovation team should hold a regular meeting focused on innovations that can bring additional value. Notwithstanding the planned meeting, if, in the course of working collaboratively to meet the goals, we discover a particular need that could be addressed by an innovative solution that would bring significant value, we may propose additional meetings.
- Prior to each meeting, a working list of candidate ideas, areas of interest, and possible services may be circulated or discussed to ensure expectations are met and needs are being appropriately addressed. These ideas may come from anyone in the team.
- These *suggestions* are part of a Potential Idea Pool that is populated and maintained based on the various interactions the innovation team has throughout the organization. Additional ideas can come from the other employees of customers, their customers, and other sources.

Innovation Log and Feedback

- The Innovation Plan includes a log of the ideas that have been considered and reviewed at the Innovation Review meetings. The Innovation Plan also tracks the lifecycle status of those that have been implemented at a summary level.
- Additionally, as part of the Innovation Plan, the innovative company solicits feedback from customers, staff, partners, even competitors via industry forums, if appropriate, on the various ideas and approaches that are presented. This information is used in conjunction with other input to refine and update the Innovation Plan and to help ensure the ongoing innovative ideas that are presented are relevant.

Early Collaboration

- Involving customers and trading partners in the innovation process can be mutually beneficial. An innovation-driven company believes in listening to its customers and giving them a voice early to help shape new products and services. For this reason, an innovation-driven company may occasionally approach its customers with opportunities to work collaboratively on new and innovative technologies, products, and services that are in the earlier stages of development.
- The scope of these initiatives will vary, but may include providing opinions, developing or evaluating requirements, providing specific feedback, or to possibly participating in trials and the early adoption of new products and services. Such a level of cooperation and collaboration helps ensure innovation is delivered as expected.

Value Challenges

A Value Challenge is the innovation equivalent of scrum techniques used for completing complex projects. Borrowing from the game of rugby, Hirotaka Takeuchi and Ikujiro Nonaka documented scrum techniques for business in a 1986 article in the *Harvard Business Review,* "New Product Development Game." Scrum is an agile development framework that is structured around sprints (or iterations of work). A key component of a sprint is the product backlog which is made up of prioritized user stories (or requirements). Scrum relies on a self-organizing, cross-functional team. The scrum team is self-organizing in that there is no overall team leader who decides which person will do which task or how a problem will be solved. Those are issues that are decided by the team as a whole. The team is cross-functional so that everyone necessary to take a feature from idea to implementation is involved. These agile development teams are supported by two specific individuals: a ScrumMaster and a product owner. The ScrumMaster can be thought of as a coach for the team, helping team members use the Scrum framework to perform at their highest level. The product owner represents the business, customers or users and guides the team toward building the right product.

Value challenges are the foundation of the Open Innovation process. However, the audience to whom a value challenge is presented may be as open or closed as the problem dictates and the problem owner prefers. While high visibility innovation tends to focus on how to create breakthrough results in an industry or be totally disruptive to an industry, value challenges tend to focus on very specific problems (generally defined in terms of an outcome not being achieved) and solving them (creating the desired outcome). A classic example of a value challenge is demonstrated by the Goldcorp Challenge, described in Chapter 7, "The Gold Standard."

Takeaway

To recap, there are several components that will be needed to execute and implement a sustainable innovation program. One of the challenges is how an innovation process is structured enough to keep the innovation process flowing, yet not so rigid as to stifle. The innovation process will need to be able to accomplish the following:

- Establish and mature innovation methodologies and common framework across the company
- Systematically identify high value external and internal targets for innovation
- Capture, evaluate, prioritize, and catalog innovative ideas
- Sponsor, develop, and refine those that have merit
- Promote the production of those that bring real value
- Facilitate adoption into operational environments
- Pilot and perform proof of concept projects
- Develop innovation experts to provide innovation consulting, training and Innovation Plan development
- Provide the processes, tools, templates, and resources that allow innovation teams to produce solid Innovation Plans

12. Innovation is a Team Sport: Play it to Win

If you consider that innovation is a team sport, and that the most important players are your customers, then innovation requires new, robust forms of collaboration with your customers. After all, inventions and other forms of innovation, such as process and business model innovations, are *in the eye of the beholder*, that is to say, in the eyes of your customers.

> Back office process improvements may save you money,
> but you cannot save your way to market leadership.
> It's your customer-facing processes that are visible and count the most.
> And that's where business intelligence and driving conversations with your customers come in.

Chicago Strategy Associates summarizes the importance of the customer in driving innovation, "Great entrepreneurs launch their businesses based on *customer insights*. In many cases they are frustrated customers. For example, a building contractor takes a family driving trip in the early 1950s and is outraged by the conditions and prices of accommodations in tourist cottages. Thus began Holiday Inn. Two lovers at Stanford grow frustrated with their email systems' inability to communicate. They seized upon a product developed by a kindred spirit in the Department of Medicine, leading eventually to commercialization of something called a router and a company named Cisco."

> The Industrial Age was about mass production.
> Innovation was R&D-driven, from the inside-out.
> It was about supply-push.
> The Customer Age is about mass customization.
> Innovation must now be driven from the outside-in.
> It's now about demand-pull.

It is about turning a company, and its entire value chain, over to the command and control of customers. This new reality demands a shift in our thinking about business innovation.

> Winning companies will be so close to their customers,
> they will be able to anticipate their needs, even before their customers do,
> and then turn to open innovation to find compelling value to meet those needs.

That, in turn, as pointed out by Drucker, means becoming a buyer for your customers. That means "business mashups" where your company joins forces with suppliers, sometimes even your competitors, to expand your product and services offerings, using the Cloud to blur industry boundaries.

Because your customers are your only true asset in the new world of low-cost suppliers, your business model will likely need to be expanded so that you can fulfill *as many of your customers' needs* as possible. This means that much innovation will be focused on business models and business processes that enable identification of needs and acquisition of ways to meet those needs. In turn, business processes and business models mean process management among organizations (internal and external) and human interaction and collaboration.

For example, the Virgin Group has established over 200 companies to meet ever more needs of its customer base. The group entered into a *coopetition* agreement with Sprint, and became the 10th largest cell phone provider in the U.S. over a short, 18-month period. Walmart is getting into banking and opening health clinics in its stores.

> Industry boundaries blur as smart companies strive
> to meet *all* they can of their customers' complete needs.

This approach is indeed a major growth strategy in a world of declining margins and commoditization. What business are you in? It had better be the *customer business!*

Now the question is, how do you get ever closer to your customers? One approach is business intelligence. Another is the Voice of the Customer (VOC). That means that now is the time to harness new means for giving your customers a voice. Back in the '90s, the voice of the customer was all the rage in business circles. It is a process discipline, a way for companies to gather customer insight to drive product and service requirements.

Techniques include focus groups, individual interviews, contextual inquiry, ethnographic techniques, etc. Each technique involves a series of structured, in-depth interviews that focus on the customers' experiences with current products or alternatives within the category under consideration. Needs statements are then extracted, organized into a more usable hierarchy, and then prioritized by the customers. Sounds logical enough.

But VOC got lost amidst the dot-com boom, abundant cheap-labor supply resources from Asia, and emerging markets as globalization reached a fever pitch. Somewhere along the way, pulling out all the stops to delight the existing customer base got lost. But today, VOC is again moving front and center, thanks to the many channels of dialog made possible by the Cloud.

Even so, customers often do not know, or cannot communicate effectively, their actual needs and requirements. Because of this, businesses need to find more creative methods of understanding customer requirements. That's why smart companies are now emphasizing business intelligence everywhere and integrating Web 2.0 communications technologies. That's why leading companies are creating blogs and wikis, placing their avatars in Second Life and taking their businesses to Facebook.

But you won't want to just open up these new Web 2.0 channels of communication and turn up the volume.

You'll want to have business intelligence embedded throughout your process management systems, and forge *meaningful collaborations* using human interaction management systems (HIMS) to tame the chaos and noise inherent in Web 2.0 technologies.

> The needed collaborations with customers
> are not one-off market research endeavors;
> they are ongoing dialogues over the lifetime of each customer.

Today's customers want it all, not just the buying transaction. Whether it is buying a PC, spare-parts, engineering services, or life insurance, customers want complete care throughout the consumption life cycle—from discovery all the way through support after the sale or contract. Today, customers demand the best deal, the best service, and solution-centered support that can only be optimized by true customer collaboration. Competing for the future is about the total customer experience, and the read-write-execute capabilities available in the Cloud are key to that experience.

Not only do companies need new means for listening to, and collaborating with customers, they also need to act on the information thus derived. To do this, they must hone the innovation process itself using the emerging science of innovation if they want to deliver ever more compelling value to their customers.

In addition to getting closer to your customers, you'll also need to get closer to your suppliers and trading partners that make up your complete value delivery system and that operate against their own clocks. As we've discussed, the days of the vertically integrated company that depends only on itself to deliver value to customers is long gone. Today it's multiple companies that participate in any given value chain and, according to generally recognized supply chain management statistics, on average over 20 companies are involved. This situation is a project manager's nightmare unless roles and responsibilities can be distributed and managed across the value chain.

Delegation is an essential part of any innovation initiative, so identifying roles and responsibilities early in a project is important. Applying the RACI (Responsible, Accountable, Consulted, and Informed) model can help. As a leader of an innovation initiative it is important that you set the expectations of people involved in your project from the outset.

Innovation initiatives require many people's involvement across many companies, so you must do all possible to avoid a situation where people are struggling against one another to do a task. Equally difficult is dealing with a situation where nobody will take ownership and make a decision. How do people know their level of responsibility; when they should involve others, or when they should exercise their own judgment?

Project Deliverable (or Activity) \ Role	C-Level Execs	Innovation Team	All Employees	Customers	Suppliers	Free Agents	Social Research Firm	Design Firm	Cloud Service Provider	Management Consultant
Discovery Phase										
Focus - Scenario Portfolio	A	R								C
Social Network Analysis	I	A	C	C	C	C	R			
Explore - Delphi	I	A								R
Incubation Phase										
Rank, Vote	I	A	C				R			
Prototype	I	A						R		
Select	A	R								
Implementation Phase										
In-house development	R/A	C						C	C	C
Co-develop	A	C		C	R	R			R	
Outsource	A	C		R	R	R			R	
License IP to others	R	C		C						
Lifecycle Phase										
Extend the Enterprise to New Markets	R									
Evolve and Adapt	A	R								
End of Life	R									

A High-Level RACI Matrix Applied to an Innovation Initiative

The RACI model is a straightforward tool used for identifying roles and responsibilities and for avoiding confusion over those roles and responsibilities during a project. The acronym RACI stands for:

- *Responsible:* The people who do the work to achieve the task. They have responsibility for getting the work done or decision made. As a rule this is one person; examples might be a business analyst, application developer or technical architect.
- *Accountable:* The person who is accountable for the correct and thorough completion of the task. This must be one person and is often the project executive or project sponsor. This is the role that the Responsible people are accountable to.
- *Consulted:* The people who provide information for the project and with whom there is two-way communication. This is usually several people, often subject matter experts.
- *Informed:* The people who are kept informed about progress and with whom there is one-way

communication. These are people that are affected by the outcome of the tasks so need to be kept up-to-date.

Without clearly defined roles and responsibilities it is easy for innovation initiatives to run into trouble. When people know exactly what is expected of them, it is easier for them to complete their work on time, within budget and to the right level of quality. A RACI matrix supports the model and is used to discuss, agree and communicate roles and responsibilities.

Indeed, innovation is a team sport involving customers, suppliers and trading partners. Choreography is essential so that each player can optimize the contribution that it makes in co-creating value.

Collaboration and the New IT Stack

We've discussed a lot about the need for collaboration in this book. But just as we dug under the covers of what the term "agility" means in Chapter 5, let's now drill down on this grand term, "collaboration." Let's turn to the article, "Riding the Fourth Wave" published in, *Information Age*, to set some context. "A new generation of people-centric collaborative information management tools is set to produce the first fundamental advances in personal productivity since the arrival of the spreadsheet. In 2002, in their seminal book *Business Process Management: The Third Wave,* Howard Smith and Peter Fingar wrote what has since come to be regarded as a manifesto for radical business change based on business process management (BPM) technology. Now though, the time is right to prepare for a new, and potentially even more radical, fourth wave of business automation, Human Interaction Management Systems (HIMS)."

"Even though much of what is described in *The Third Wave* has still to be realized, among its most sophisticated early adopters, BPM has already eliminated most of the back-end system bottlenecks that have traditionally impeded business development. For these organizations, it is time to move on. The real future, if you look at business process management, the key part of it that has not been fully addressed is human-to-human interaction."

"To some extent, this assertion is already recognized in the current industry vogue for collaborative, Internet-based personal productivity tools such as Google's word processor and spreadsheet products. Unlike first generation Microsoft Office-like applications, such so-called Office 2.0 products are designed from the ground up to distribute and share documents. However, HIMS proponents believe that these advances do not really solve human interactivity problems, and may actually be making them far worse [e.g. with *infoglut*]."

What Smith and Fingar wrote about in their book didn't in any way ignore human interaction management. The people components of a business process were given equal status to machine components in their definition of a business process, "A business process is the complete and dynamically coordinated set of *collaborative* and *transactional* activities that deliver value to customers." But the IT industry went its own way, by and large focusing on the "transactional activities," relegating the collaborative part to be trapped in the world of traditional workflow, with BPM lipstick. BPM vendors with a strong workflow heritage began labeling their BPM suites as "people-oriented." Meanwhile, other vendors competed in the space known as "integration-centric BPM"—hmm, a new term for next-generation Enterprise Application Integration (EAI)?

Such labeling is a red herring, for the way humans interact among themselves to get work done, especially in the world of business innovation, is far different from integration-centric BPM or the predefined notion of workflow, even with complex nesting and chaining logic built in. These are primarily notions of system-to-human (S-2-H) systems, where people are treated as cogs in an assembly line, dynamic as it may be, shoving tasks from station to station. The chairman of the Workflow Management Coalition (WfMC), Jon Pyke, noted, "Supposing you were playing golf; using the BPM approach would be like hitting a hole in one every time you tee off. Impressive – 18 shots, and a round finished in 25 minutes. But as we all know, the reality is somewhat different (well, my golf is different) – there's a lot that happens between teeing off and finishing a hole. Ideally, about four shots (think nodes in a process) – but you have to deal with the unexpected even though you know the unexpected is very likely; sand traps, water hazards, lost balls, free drops, collaboration with fellow players, unexpected consultation with the referee – and so it goes. Then there are 17 more holes to do – the result is an intricate and complex process with 18 targets but about 72 operations."

Like the game of golf, when it comes to the *creative* and *innovative* forms of business processes that reside in the domain of human-to-human interactions, the processes cannot be predefined or "flowcharted" in advance. In short, such collaborative human processes are "organic." That is to say, as suggested above, they represent "emergent processes" that change not only their state, but also their structure as they are born, and then grow and evolve. Such processes deal with new business initiatives, new programs, new marketing campaigns, new product development, case management, research, and all too often, unexpected crises. These are not the kind of processes you call in IT to analyze, model and code – and get back to you in 18 months with a solution. But these are the kind of processes at the heart of business innovation.

In the messy real world of business, people communicate, research, think, consult, negotiate and ultimately commit to the next steps that are unknowable at the outset. As new commitments are made, the process continues, often involving new participants playing new roles as the process expands. The participants usually cross organization and company boundaries: functional departments, customers, regulatory agencies, suppliers, suppliers' suppliers, design firms, market research firms, channel partners, and so on. Unlike the internal command and control within a single company, one company cannot command another company to do this or that. Further, each participant operates using its own time clock.

Instead, the parties must negotiate and commit to next steps, and track the many agreements made along the way. Such human collaboration shifts the requirements for IT support from "information processing" where data are tracked, to "commitment processing" where agreements are tracked. Does your ERP or BPM system do that?

Human Communication with Implicit Collaboration

Let's take a brief look at the tools people use to carry out knowledge work, decision-making and collaboration. It's not a pretty picture, but we need to grasp the diverse and complex nature of how humans interact to accomplish their goals, as shown below. Portrayed in the figure are people from four companies:

- *Company J* (Joe, Jane, and John),
- *Company T* (Tim, Terry, Tanya, and Tom),
- *Company G* (Greg and Gina) and
- *Company C* (Carl, Casey, and Cassie).

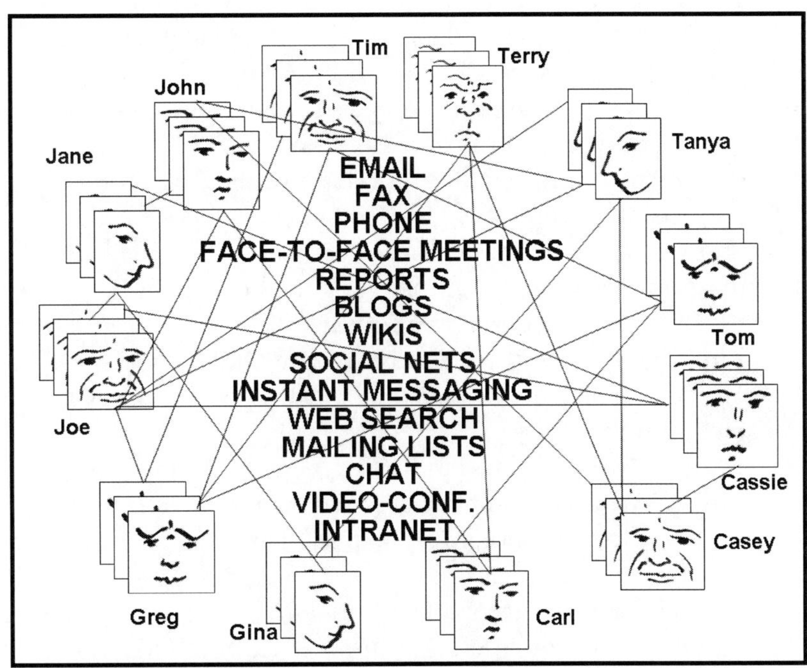

Communication With "Implied" Collaboration (A Real Mess)

☑ *Company J* wants to develop a new financial service to provide a new form of health insurance that allows consumers to utilize the growing number of low-cost, high-quality health providers (hospitals, surgeons, assisted living facilities, and nursing homes) in the emerging Globalized Health Care industry. Joe is the CEO, Jane is the V.P. of Business Development, and John is an M.D. with international health expertise.

☑ *Company T* is a health services company representing transnational hospitals and related organizations based in India, Thailand, and Israel. Tim is V. P. of Business Development, Terry is V.P. of Quality and Government Compliance, Tanya is Research Director, and Tom is V.P of Finance.

☑ *Company C* is a highly specialized broker responsible for channel development. Cassie is the CEO and Casey is V.P of Business Development.

☑ *Company G* is an international law firm located in The Hague that specializes in international medicine regulations and the intricacies of related global trade agreements.

Somewhere in all their communications, and all their "Web 2.0" endeavors, there must be collaboration taking place, right?

Of course there is, but it's "implicit collaboration,"—keeping track of what's really going on is all in their heads, each head having its own assessment of what's going on at any particular time.

Yet there is no real technology support to structure their collaborations. Quite the contrary, the participants suffer from information overload, also known as infoglut. Who got "cc:'ed" on the latest version of the risk assessment projections? Who have or have not completed their latest critical path tasks? Which experts do we now need to bring to the table? Was the contract signed with the Thai government? Do we have a business plan for acquiring and managing the hospital in Belize? Now that the contract has been signed for managing the hospital in Panama City, what are the next steps, and who and what are needed to move forward? Where's the latest version of the lobby layout? Have the investors committed to the Belize project? What steps do we still need to complete for certifying our Indian medical staff in Costa Rica? Did we get responses to our RFP from the pharmaceutical companies in India?

Such is the real world of new business initiatives, new marketing campaigns, new joint ventures, mergers and acquisitions, research, and business innovation in general. It's a world of human interactions that to date have little technological support that truly provides "human interaction management."

Enterprise software companies have made moves in recent years to offer human communication facilities as a part of their software offerings. Through the combined power of social software and unified communications, the connections between people, information, and communities become visible and actionable. While such offerings are a great step forward in providing communication and implicit collaboration, something essential is missing when it comes to goal-oriented, explicit collaboration.

Earlier, we talked about open innovation, casting a wide net, crowd sourcing, and collective intelligence. These all play a vital role in the OO part of our OODA loops. But what about the Decide and Act parts? Is there a need for "closed innovation?" Is there a need to shift gears from weak ties for gaining insight to strong ties for actually doing the work of implementing an innovation? Let's turn to a father of virtual reality and author of *You Are Not A Gadget*, Jaron Lanier.

tinyurl.com/84vk6pc

"The problem is not inherent in the Internet or the Web. Deterioration only began around the turn of the century with the rise of so-called "Web 2.0" designs. These designs valued the information content of the web over individuals. It became fashionable to aggregate the expressions of people into dehumanized data. There are so many things wrong with this that it takes a whole book to summarize them. Here's just one problem: It screws the middle class. Only the aggregator (like Google, for instance) gets rich, while the actual producers of content get poor. This is why newspapers are dying. It might sound like it is only a problem for creative people, like musicians or writers, but eventually it will be a problem for everyone. When robots can repair roads someday, will people have jobs programming those robots, or will the human programmers be so aggregated that they essentially work for free, like today's recording musicians? Web 2.0 is a formula to kill the middle class and undo centuries of social progress.

"On one level, the Internet has become anti-intellectual because Web 2.0 collectivism has killed the individual voice. It is increasingly disheartening to write about any topic in depth these days, because people will only read what the first link from a search engine directs them to, and that will typically be the collective expression of the Wikipedia. Or, if the issue is contentious, people will congregate into partisan online bubbles [digital Balkanization] in which their views are reinforced. I don't think a collective voice can be effective for many topics, such as history—and neither can a partisan mob. Collectives have a power to distort history in a way that damages minority viewpoints and calcifies the art of interpretation. Only the quirkiness of considered individual expression can cut through the nonsense of the mob—and that is the reason intellectual activity is important.

"On another level, when someone does try to be expressive in a collective, Web 2.0 context, she must prioritize standing out from the crowd. To do anything else is to be invisible. Therefore, people become artificially caustic, flattering, or otherwise manipulative.

"Web 2.0 adherents might respond by claiming that I have confused individual expression with intellectual achievement. This is where we find our greatest point of disagreement. I am amazed by the power of the collective to enthrall people to the point of blindness. Collectivists adore a computer operating system called LINUX, for instance, but it is really only one example of a descendant of a 1970s technology called UNIX. If it weren't produced by a collective, there would be nothing remarkable about it at all.

"Meanwhile, the truly remarkable designs that couldn't have existed 30 years ago, like the iPhone, all come out of "closed" shops where individuals create something and polish it before it is released to the public. Collectivists confuse ideology with achievement.

"There are some cases where a group of people can do a better job of solving certain kinds of problems than individuals. One example is setting a price in a marketplace. Another example is an election process to choose a politician. All such examples involve what can be called optimization, where the concerns of many individuals are reconciled. There are other cases that involve creativity and imagination. A crowd process generally fails in these cases. The phrase "Design by Committee" is treated as derogatory for good reason. *That is why a collective of programmers can copy UNIX but cannot invent the iPhone.*

"In the book, I go into considerably more detail about the differences between the two types of problem solving. Creativity requires periodic, temporary 'encapsulation' as opposed to the kind of constant global openness suggested by the slogan 'information wants to be free.' *Biological cells have walls, academics employ temporary secrecy before they publish, and real authors with real voices might want to polish a text before releasing it.* In all these cases, *encapsulation* is what allows for the possibility of testing and feedback that enables a quest for excellence. To be constantly diffused in a global mush is to embrace mundanity."

So, what's needed to move on to "Doing the Work," to being goal-oriented in getting the innovation job done?" That would be *explicit collaboration.*

Explicit Collaboration Via Human Interaction Management Systems

If technology is to be used to support human interactions for innovation, collaboration can no longer be *implicit.* It must be *explicit* if it is to be brought under management control, not the Web 2.0 mob Lanier writes about. For this to happen, 5 basic principles are needed:

1. *Connection visibility:* to work with people, you need to know who they are, what they can do, and what their responsibilities are as opposed to yours.
2. *Structured messaging:* if people are to manage their interactions with others better, their communications must be structured and goal-directed.
3. *Support for knowledge work:* organizations must learn to manage the time and mental effort their staff invest in researching, comparing, considering, deciding, and generally turning information into knowledge and ideas.
4. *Supportive rather than prescriptive activity management:* humans do not sequence their activities in the manner of a procedural computer program. There is always structure to human work, sometimes less and sometime more, but it is not the same kind of structure that you get in a flowchart.
5. *Processes that change processes:* human activities are often concerned with solving problems, or making something happen. Such activities routinely start in the same fashion – by establishing a way of proceeding. Before you can design your new widget, or develop your marketing plan, you need to work out how you are going to do so – which methodology to use, which tools to use, which people should be consulted, and so on. In other words, process definition is an intrinsic part of the process itself. Further, this is not a one-time event – it happens continually throughout the life of the process. Human interaction management requires a major shift from "information processing" to "commitment processing," where participants negotiate and commit to next steps. The process itself is *emergent*, not predefined.

To achieve all this, a new kind of software system is required, one based on the six different kinds of "objects" defined by Human Interaction Management (HIM): Roles, Users, Interactions, Entities, States and Activities. This book is not the place to discuss the nature of each HIM object type, but all six must be used as the fundamental basis of any system intended to properly support collaborative human work in the enterprise.

By implementing these principles in software, a human interaction management system can support human collaboration in a way that can effectively turn strategy into action, and provide the mechanisms for ensuring strategic, executive and operational management control. All these forms of management control have direct ramifications for corporate and government compliance. Implementing these principles allows management participation in the process execution, including ongoing re-definition of the process itself, thereby ensuring maximum *agility* and *responsiveness*.

With support from a Human Interaction Management System (HIMS), an organization can provide new means to help people work better as they take on the constant stream of new business initiatives and the human-centric tasks that make successful companies tick:

☑ new product development,
☑ promotions and special events,
☑ research,
☑ new marketing campaigns,
☑ customer self-service,
☑ case management,
☑ mergers and acquisitions,
☑ opening new global markets,
☑ complex sales proposals,
☑ management-level Sarbanes-Oxley (SOX) act compliance,
☑ the projects undertaken to innovate, to grow the business, or
☑ to stave off competitive threats and deal with exceptions and crises.

This is the stuff people do that allows companies to go from good to great. It's about going beyond efficiency and on to effectiveness, going beyond being a commodity player to becoming an innovator. It's about dealing with tacit information, not transactional data.

The figure below illustrates the use of a HIMS to support human work. Each of the participants, typically scattered across companies as we previously discussed, plays *multiple roles* in *multiple projects*. Such roles can vary from "responsibility" for a given project to simply "being informed" on other projects.

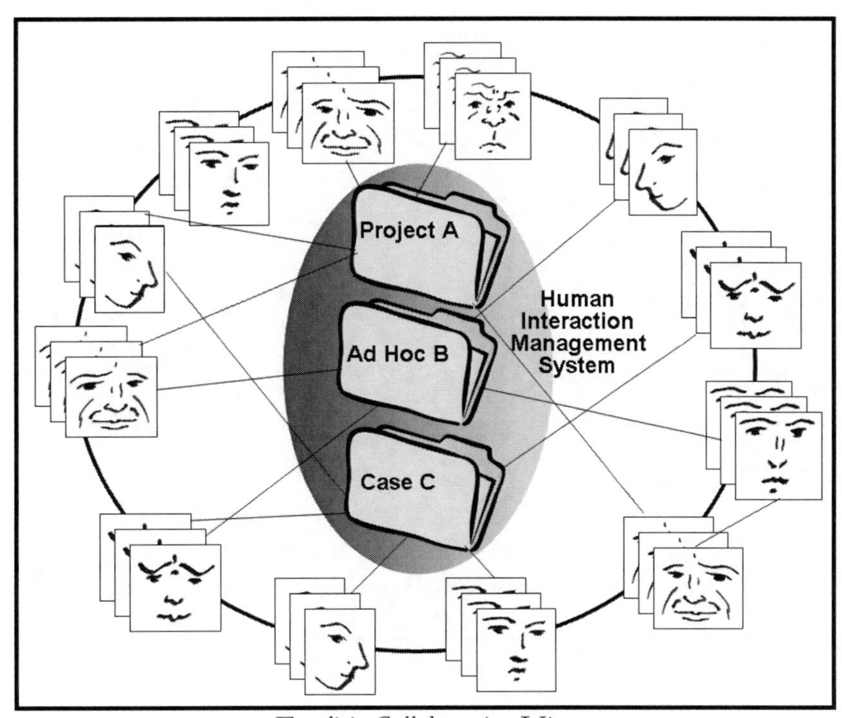

Explicit Collaboration Via a
Human Interaction Management System

The HIMS is used to add *structure* to collaborations, making them *explicit*. Collaborations can be structured around specific *goal-driven* projects or cases, but can also be used to add structure to ad-hoc collaboration, taming what is now "in-box hell." Just consider the "global CC: nightmare" wherein people just CC:/FWD: emails to each other randomly, rather than taking proper responsibility for sorting out issues. According to a BBC report in the UK, it's not unusual for office workers to spend as much as two hours a day, every day, sorting and reading all the email that pours into their in-boxes. Worse, that doesn't include the time they have to spend responding to it.

Let's not forget that today's companies are complex and depend on their IT systems. Thus we can now take the figure and surround the people involved in human collaboration with the "background" IT systems that are embedded in most organizations, as shown below.

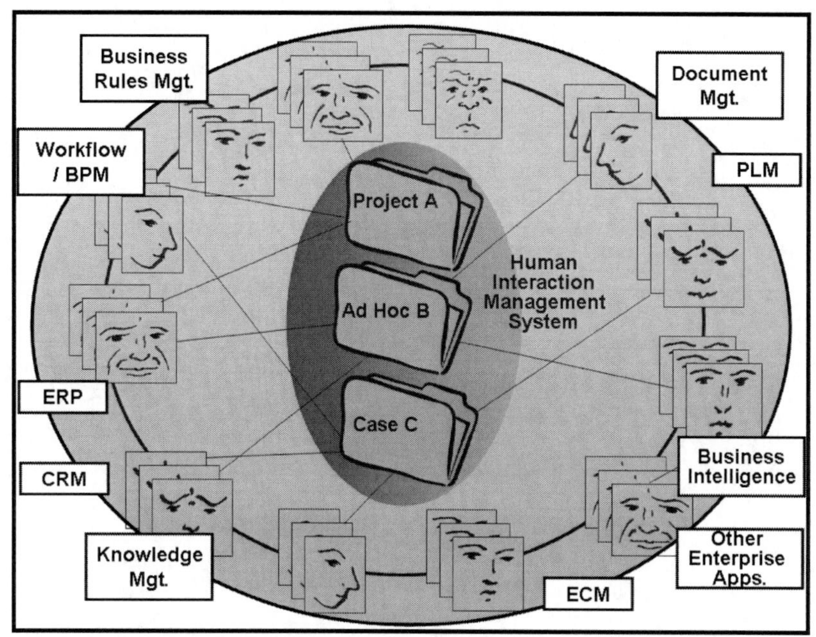

Modern Businesses Depend on Multiple and Disparate IT Systems

To many in a typical organization, they have to get their thinking and decision-making work done in the context of, and often while maintaining, a host of often disparate enterprise systems as illustrated in the figure. Such is the case for, let's say, a sales manager who must do the real work of collaborating with customers, suppliers and sales people, while also feeding and caring for the CRM machine. But because, as Forrester Research points out, 85% of all business processes involve people, and because human-driven processes have not been fully addressed in the majority of today's BPM systems, we simply started with people in this discussion. After all, as John Seely Brown once pointed out, "processes don't do work, people do." There are essentially two ways people are involved in business processes:

1. As cogs in a workflow where they are asked to "approve" or "not approve" predefined next-steps, or "do this task" or "post this transaction," or "escalate this problem;" and
2. As knowledge workers and decision makers going where no routine, transactional process has gone before: not simply operating today's business, but charting tomorrow's course, or addressing major disruptions in the industry.

The first case, cogs in a workflow, is handled reasonably well at present by workflow systems—except, that is, for the 20% of *special cases* that actually consume 80% of your resources! This is of deep concern to telcos, for instance, who are struggling desperately to control the costs of resolving line faults that cannot be dealt with automatically and thus go to manual "case management" mode – costing some telcos millions of dollars.

In the second case, the knowledge worker and decision maker has little support whatsoever from workplace applications at present. What they need is a new kind of interface to their existing tools that allows automated integration of knowledge work with routine business processes handled by heritage workflow and ERP systems.

Yes, people are a part of almost all processes, excepting straight thru processes (STP) such as a financial trade or other routine transactions. It's the "non-routine" that drives competitive advantage, and that's what human-driven business processes are all about. Indeed, shouldn't this class of process be paramount in any business?

After all, the routine, operational processes are commodities, and who wants to be a commodity player in today's hypercompetitive global marketplace?

Michael Hugos writes in his book, *The Greatest Innovation Since the Assembly Line,* that companies should be "efficient enough," but the real prize is responsiveness. Being responsive to changing market conditions, competitive threats and new market opportunities is what successful business is all about. And, that's all about people and how they collaborate to innovate.

We still need continuous process improvement via workflow and integration-style BPM, and many companies are just now coming to grips with these process-oriented IT systems. Service-oriented architecture (SOA) has enabled great advances and is currently all the rage. But SOA, by itself isn't enough. BPM pioneer, Ismael Ghalimi, said it best, "BPM is SOA's killer application, and SOA is BPM's enabling infrastructure." Human interaction management pioneer, Keith Harrison-Broninski takes that notion one giant step forward, "HIM is BPM's killer application."

SOA is primarily in the domain of IT analysts. BPM is primarily in the domain of business analysts. HIM is primarily in the domain of business people themselves, providing support for the way humans actually work and interact with one another.

The next figure ties together the world of human interactions with the surrounding world of IT systems. The HIMS puts heritage IT systems in their proper place, for humans don't work together in isolation from their IT Systems (even though many wish they could).

The HIMS is not a replacement for workflow and BPM any more than business rules management systems (BRMS) or a business intelligence systems (BI) are. They are orthogonal. They are complex systems in their own right, and in their own domains.

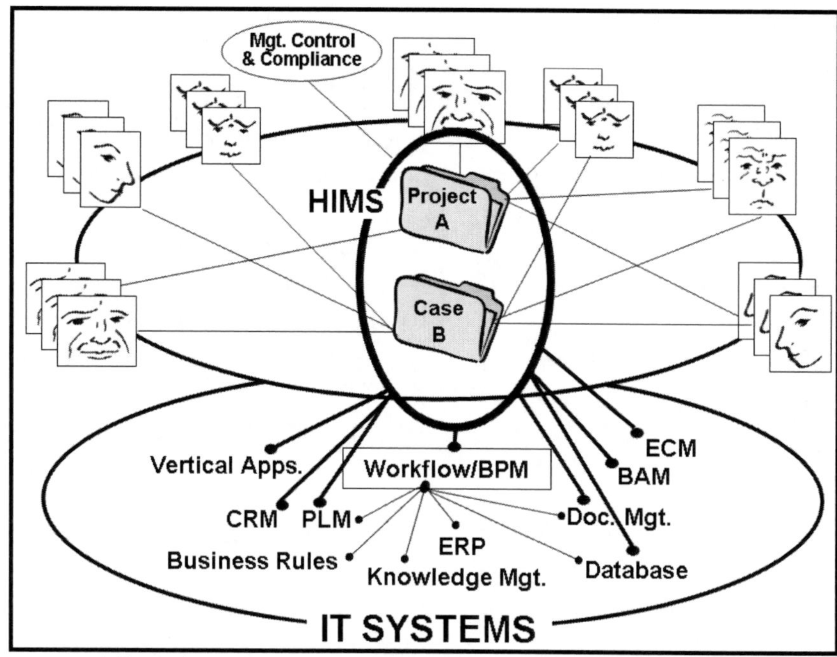

The Relationship Between the HIMS and Heritage IT Systems

But just as a BPM system can fuse disparate IT systems into end-to-end business processes (esp. with the help of Web services and SOA), the HIMS must have the capability to fuse workflow/BPM and other heritage systems into the "information base" knowledge workers use when they collaborate via human-driven processes.

The HIMS can help keep non-routine work activities in context (telling everyone "what's going on,") and can change the way executives, line managers and high-value knowledge workers manage their work as they guide the business into the future.

But for all this to come about, the design of an enterprise-class HIMS must incorporate the underlying computer and social sciences that reflect how humans actually interact together. They must have the underpinnings of:

☑ negotiate-and-commit speech acts, so that agreements, not just information can be tracked, and so that human-driven processes can be redefined as they emerge/evolve over the lifetime of the process,
☑ role activity theory, where human interactions, not computer interactions, can best be modeled,
☑ distributed computing techniques to cope with multiple and dynamic asynchronous communication channels, allowing a given human-driven process to be redefined as participants in that process are dropped in or out (Mobile Processes based on Pi Calculus),

☑ multi-agent systems, where collaborating software agents use their own unique business rules and knowledge sources,
☑ cognitive science models (such as REACT/AIM) that reflect how knowledge workers think and act,
☑ choreography methods for handling interactions with process participants, including stable, operational processes in heritage IT/BPM/Workflow systems, and
☑ private information spaces, where versioning and shared access are under the control of the participants that own the information, work objects and processes.

These underpinnings should never be seen by business users of such systems, just as relational algebra isn't seen by business users of modern ERP systems. But they had better be there, or else the HIMS will not be flexible enough for adaptive and dynamic collaborations in which the interaction patterns cannot be anticipated. The activity sequencing of human interactions cannot be prescriptively imposed. Why? The contracts of interactions, deliverables and business rules are continually renegotiated during the life of the process: emergent behavior, emergent processes. The HIMS pushes the envelop of the typical workflow and integration approaches in managing these kinds of "impromptu" processes.

So, let's put all this together.

Today, there is a need to come to grips with support for human interaction management for non-routine, knowledge-based workplace activities, and to address the desperation people feel when they are swamped by increasing demands and infoglut in today's workplace. In "Linking Insight to Action: The Next Big Goal," posted at *Intelligent Enterprise*, Doug Henschen reveals a real problem in business, "I've had a number of conversations in recent days around the theme of linking analysis to action. There's lots of frustration out there, understandably, because managers and executives increasingly have plenty of tools that spot problems – reports, alerts, dashboards, KPIs, scorecards, etc. – but they're not connected to levers that enable them to take action."

To reach for Doug's next big goal, we can turn to EDS's Janne Korhonen's depiction of the evolution of process management as shown in Figure 5 and relate that to how HIM can turn strategy into action:
☑ *Strategic Control:* the definition of aims and measures for each high-level process
☑ *Executive Control:* the definition of outline processes that include a mixture of Roles, Interactions and Users
☑ *Management Control:* adapting the outline processes into a form for initial execution and later ongoing redefinition of the process, along with executive feedback (e.g. dashboards, statistical reports)
☑ *Agreements:* contract of interactions, deliverables and business rules is continually renegotiated during the life of the process

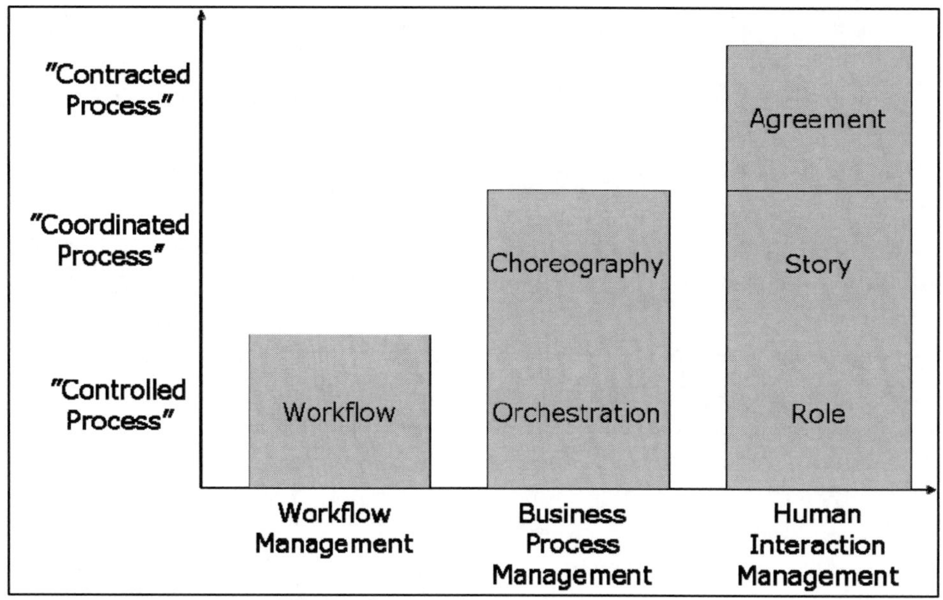

The Evolution of Process Management
(Source: www.jannekorhonen.fi/blog/wp-content/BPM_Systemic_Perspective.pdf)

Takeaway

To paraphrase Admiral David Farragut, "Damn the flowcharts, full business speed ahead." Flowchart-based technologies treat work as steps in a routinized process, and hence people as cogs in a machine. Because such technologies don't "understand" knowledge work, they can only mimic its superficial appearance as a sequence of tasks. Tasks are but the tip of the knowledge-work iceberg. It's time to take insight, innovation and strategy into action. That's about people and dynamic human interactions, not predefined workflow/BPM.

- It's time to go beyond assembly-line workflow.
- It's time to filter out the noise of social network chatter.
- It's time for explicit collaboration.
- It's time for human interaction management.
- It's time for a new IT stack.

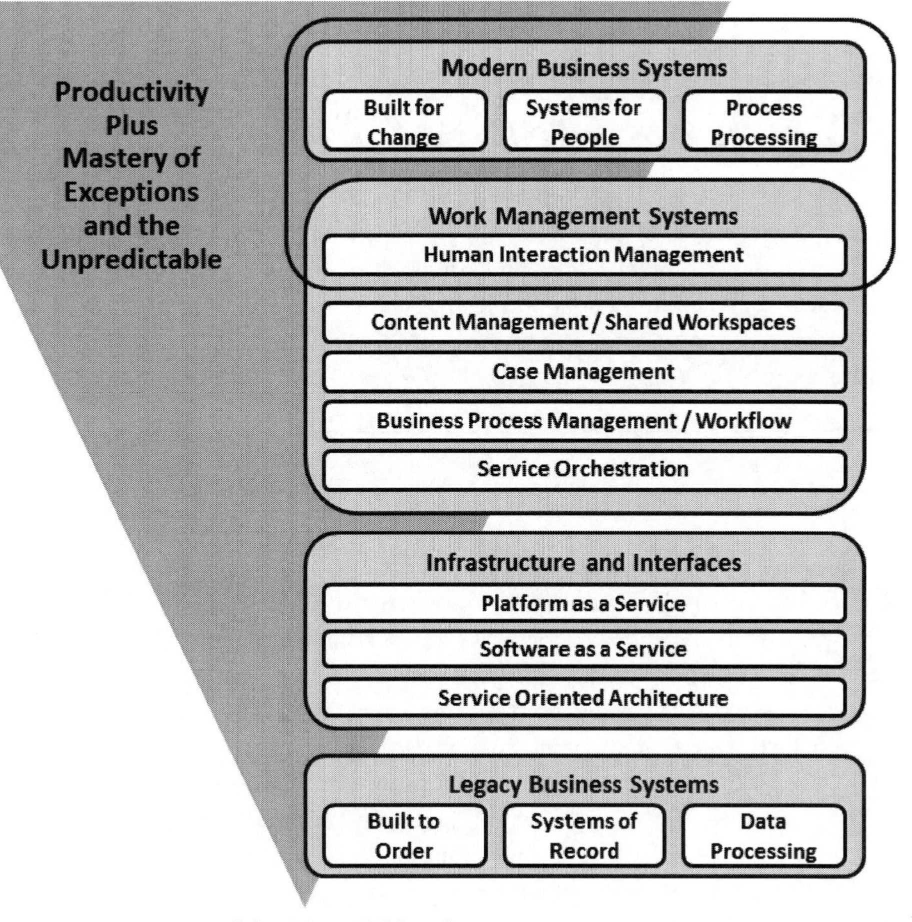

The New IT Stack

13. Big Data and Predictive Analytics

"Big data—large pools of data that can be captured, communicated, aggregated, stored, and analyzed—is now part of every sector and function of the global economy. Like other essential factors of production such as hard assets and human capital, it is increasingly the case that much of modern economic activity, innovation, and growth simply couldn't take place without big data."—McKinsey Global Institute, *"Big data: The Next Frontier for Innovation, Competition, and Productivity,"* June 2011.

Big Data + Predictive Analytics =
Actionable Business Insight/Foresight

Driven by "The Internet of Things (IOT)," in the era of "Big Data," billions of devices connect to the Web and create a continuous stream of data on the actions, locations, and conditions of everything from livestock to people—does your pet have a chip yet? According to Cisco, the number of devices connected to the Web exceeded the number of people connected in 2008. A near invisible network of radio frequency identification tags (RFID) is being deployed on almost every type of consumer item or industrial part. These tiny, traceable chips, which can be scanned wirelessly, are being produced in their billions and are capable of being connected to the internet in an instant. This so-called "Ambient intelligence" promises to create a global network of physical objects every bit as pervasive and ubiquitous as the World Wide Web itself.

In short, data have swept into every industry and business function and are now an important factor of production, alongside labor and capital. Charles L Mauro, Founder of MauroNewMedia recently concluded that "the social life of devices" will be just as important as the social life of those who use them. By this he means that how devices behave in the new universe of hyperconnectedness will determine success from both a business and personal perspective.

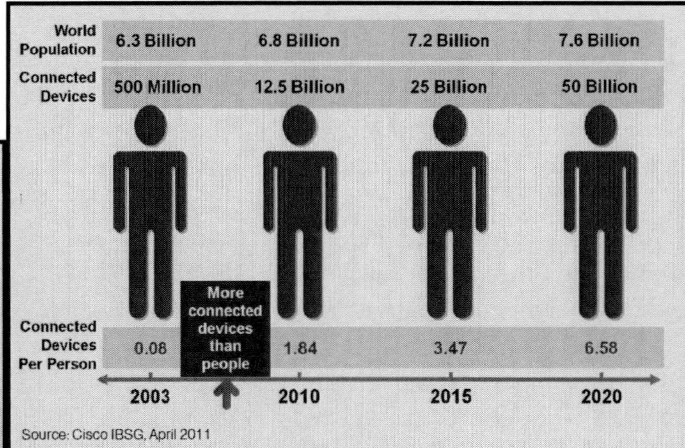

Source: Cisco

Watch: The Internet of Things by IBM
tinyurl.com/yktzhqs

The IT analyst firm, Gartner, provides a description of big data, "Big data refers to the volume, variety and velocity of structured and unstructured data pouring through networks into processors and storage devices, along with the conversion of such data into business advice for enterprises." These elements can be broken down into three distinct categories: *volume, variety and velocity:*

- *Volume (petabytes and eventually exabytes):* The increasing amount of business data—created by both humans and machines—is putting a major hit on IT systems, which are struggling to store, secure and make accessible all that information for future use.
- *Variety:* Big data is also about the increasing number of data types that need to be handled differently from simple email, data logs and credit card records. These include sensor- and other machine-gathered data for scientific studies, health care records, financial data and rich media: photos, graphic presentations, music, audio and video.
- *Velocity:* It's about the speed at which this data moves from *endpoints* into processing and storage.

Nice definition, but what does Big Data really mean as far as it's business impact? Ken Rosen, Managing Partner at Performance Works isn't the least bit happy with Gartner's definition as he thinks it misses the real point, "That's like saying 'New ideas come from electricity moving among brain cells.' It's correct, but emphasis is wrong. It makes sense that an IT-oriented firm like Gartner would focus on speeds, feeds, and infrastructure, but executives need a different view.

"Let's be clear, Big Data is *not* simply dealing with lots of data! I sympathize with colleagues who hate the name. After all, 1,000 movies is a Petabyte. But 1,000 movies is not a Big Data problem. On the other hand, I just spoke with the CTO of a top Pharmaceutical firm. They have a serious Big Data initiative, and the total data fits on a single hard drive.

"So what IS Big Data? *New meaning from new sources.* Big Data is finding new meaning from new data sources. New meaning that was never practical to find before—because of scale, data format, distribution of data in many locations, the fact that no one thought of looking before, etc. Sources from Lego store purchase patterns to iPhone GPS information. From automobile traffic patterns to Internet data traffic patterns. From weather patterns to earthquakes. From tech support response times to medication response times. It is *easily* as much a new mindset as new technology. Again, *new meaning from new sources.*

"Why should you care? Because just for a start, businesses can learn what to offer and to whom. When to offer something new and through what channels. Which employee can best solve a problem and when to get outside help. Which competitor will win and when their stock price will reflect the victory. I'll go out on a limb here: I consider Big Data the most important thing for business since the Internet."

Indeed Big Data is all about delivering new insights to decision makers. Here's an example. As reported in a *Forbes* article, Walmart wanted to find out the biggest-selling items people bought before a hurricane hits. The No. 1 answer—batteries— was not a surprise. But the unexpected No. 2 item was Kellogg's Pop-Tarts. They last a long time, don't require refrigeration or preparation, and are easy to carry and store. As a result of this intelligence, Walmart can now stock up on Pop-Tarts in its Gulf Coast stores ahead of storm season. This is where the reach of new-generation business analytics tools shine by directly helping enterprises make smart decisions.

Historically, data analytics software hasn't had the capability to take a large data set and use it to compile a complete analysis for a query. Instead, it has relied on representative samplings, or subsets, of the information to render results. That approach is changing with the emergence of new big data crunching engines, such as the open-source Apache Hadoop. Hadoop and other such systems provide complete looks at big data sets. Instead of a team of analysts spending days or weeks preparing the parameters for data subsets, and then taking 1, 2 or 10 percent samplings, all the data can be analyzed at one time, in real time. Why bother? Because data sitting in storage arrays and Cloud accounts represents unrefined value in its most basic form. If interpreted properly, the stories, guidelines and essential information buried in storage and databases can open the eyes of business executives as they make strategic decisions for their company.

Apache Hadoop, open-source software, has proved to be the data prospector with the most market traction in the last several years. Hadoop processes large caches of data by breaking them into smaller, more accessible batches and distributing them to multiple servers to analyze. It's like cutting your food into smaller pieces for easier consumption. Hadoop then processes queries and delivers the requested results in far less time than old-school analytics software—most often minutes instead of hours or days.

Doug Cutting, Apache Foundation Chairman, named his new creation Hadoop, after his son's big stuffed elephant.

IBM, the first large systems maker to use the engine, provides its Hadoop-based InfoSphere BigInsights. CEO Sam Palmisano revealed in an August 2011 presentation, "In about a year from now, you'll be starting to see the fruits of our 'big bet' on big data. The work we've been doing for the last several years with Watson [the IBM computer that won Jeopardy! matches against two human champions] will move into products that will be used for a great many purposes, including health care, science and financial applications. Our engineers say they're not far away from building a supercomputer about the size of a human brain that can fit into a shoebox."

As a technology aside, we'll need a big "refresh" in the underlying architecture of the Internet if we are to handle the exponential growth in the velocity, variety and volume of Big Data. A key to the next-generation architecture centers on *Content-Centric Networks,* "The philosophy behind content-centric networks was pioneered by Ted Nelsonin 1979 and later by Brent Baccala in 2002. In 1999, the TRIAD project at Stanford proposed avoiding DNS lookups by using the *name* of an object to route towards a close replica of it. In 2006, the DONA project at UC Berkeley and ICSI proposed a content-centric network architecture, which improved TRIAD by incorporating security (authenticity) and persistence as first-class primitives in the architecture. In 2009, PARC announced their content-centric architecture within the CCNx project, which is led by Van Jacobson, a research fellow at PARC. On September 21, 2009, PARC published the specifications for interoperability and released an initial open source implementation (under GPL) of the Content Centric Networking research project on the Project CCNx site (ccnx.org)."

Watch Jacobson: *tinyurl.com/6s5lfer*

What IBM is talking about is Big Data "predictive analytics." The term "business intelligence (BI)" was coined way back in 1958 and companies have relied on BI ever since. But most of the results of BI only show *what has happened,* not *what's most likely to happen next.* That's where predictive analytics come in.

Predictive analytics is an area of BI that deals with extracting information from data and using it to *predict* future trends and behavior patterns. The core of predictive analytics relies on capturing relationships between explanatory variables and the predicted variables from past occurrences, and exploiting it to predict future outcomes. While traditional BI can tell us where we've been, predictive analytics can tell us where we are likely to be going so that more rapid and insightful decisions can be made.

The term predictive analytics is used to mean predictive modeling, "scoring" data with predictive models, and forecasting. *Predictive models* analyze past performance to assess how likely a customer is to exhibit a specific behavior in the future in order to improve marketing effectiveness. Amazon serves up an example of predictive models that most of us can relate to when we visit its site and it recommends what we might like based on our individual buying history and patterns.

Decision models, on the other hand, describe the relationship between all the elements of a decision—the known data (including results of predictive models), the decision and the forecast results of the decision—in order to predict the results of decisions involving many variables. These models can be used in optimization, maximizing certain outcomes while minimizing others. Decision models are generally used to develop decision logic or a set of business rules that will produce the desired action for every customer or circumstance.

Deep analytics may use several quantitative methods broadly grouped into regression techniques and machine learning techniques. These methods can apply to many industries, including financial services, insurance, telecommunications, retail, travel, healthcare, and pharmaceuticals.

The book, *Competing on Analytics,* serves up many stories of exemplars other companies can learn from in their pursuit of competitive advantage through next-generation analytics.

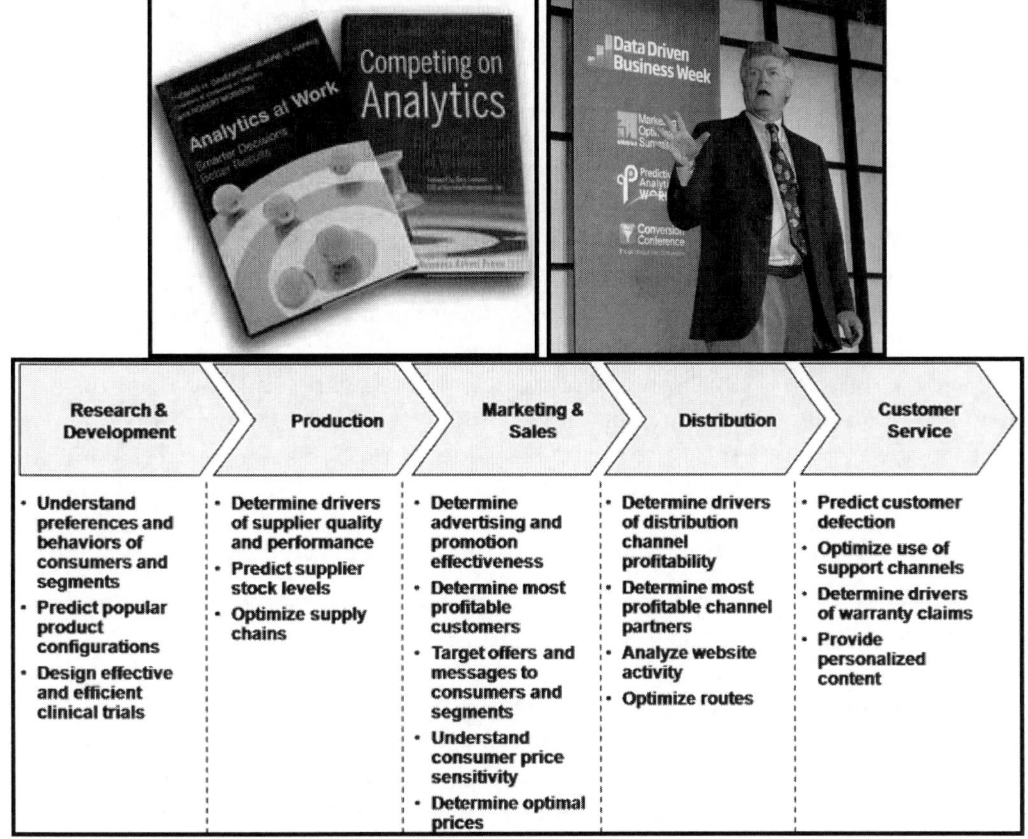

Tom Davenport, Jeanne Harris and Robert Morison

Never mind 20th century focus groups and marketing research surveys. Let the Big Data + Predictive Analytics games begin, for they are not just about computing and databases, they are about *a new generation of analytics driving business insights for business innovation in the Cloud.*

Of course, to act on this new insight requires mastery of business process management, for BPM is how stuff actually gets done in light of new insights.

Forget BPMN swim lanes going from as-is to to-be.

Start with "What the hell for?"

That's what big data with deep analytics is all about—showing me the patterns and trends that I couldn't otherwise begin to fathom.

Acquire what IBM-emeritus guru James Martin, calls "alien intelligence" that only deep analytics across big data can provide, then you can develop dynamic to-be processes that can change in flight, in real-time, to cope with today's new world of uncertainty.

> If BPM was an airplane, Big Data would drive its navigation system, in real time, just as it already drives the navigation of jumbo jets across continents and oceans.

There are several ways Big Data can support innovation. It can make information transparent and usable more often and across a broader audience supporting collaboration. As more real-time data is available, more accurate and detailed performance information is available for observation and insight. It can be used to conduct controlled experiments to make better management decisions. Larger data sets allow far greater means to segment customers and therefore much more precisely tailored products or services (mass customization).

Larger data sets and sophisticated analytics can substantially improve decision-making and risk management of an innovation portfolio. Total product, service, customer life cycle data can be used to improve the development of the next generation of products and services—including add-ons and support offerings such as sensor data (cars calling home), warranty and repair work, customer inquiries and return purchases.

Embedded real-time Big Data analytics will be critical in supplying the contextual information that future generation services need to compete—such as location enabled applications and consumer intent based upon past (and immediate past) behavior and requests.

There is, however, also a risk of over analysis with Big Data with convergent analytical thinking driving out divergent, creative, and intuitive thinking.

Let's consider Observation and Orientation and their relation to Insight and Foresight. When we discussed an innovation architecture, we talked about the Observation and Orientation elements of OODA. *Insight* comes from the trends, values, summaries of the information that can be gleaned from Big Data indicating what is happening *right now*. Foresight comes from the anomalies, and weak signals (blips in the data) indicating potential changes that are coming.

Big Data is a way to engage customer collaboration in co-creation of value, without bothering them. Customers become passive business consultants for companies via their purchases, searches, and online activities being tracked to improve everything from Web sites to delivery routes. Mobile phones, automobiles, factory automation systems and other devices are routinely enabled with RFID capabilities to generate streams of data on their activities, making possible an emerging field of "reality mining" to analyze this information and to identify white space opportunities for new innovations.

When it comes to new frontiers in decision making, predicting the market performance of an innovative idea requires an understanding of the difference between data and knowledge. More importantly, an innovation manager must plan how to meet customers' needs based on their behaviors. This becomes straightforward through the use of predictive analytics, and the yield of successful innovative products in a balanced portfolio becomes much more predictable no matter which innovation development processes are ultimately engaged. This is beyond the sensing, Observation and Orientation of OODA, and into the decision-making process, e.g., how to allocate resources among all those new ideas.

Science has progressed to where we are today via experimentation. Running controlled experiments is the gold standard for sorting out cause and effect. But experimentation has been difficult for businesses because of cost, speed and convenience. Now the Cloud will allow business to conduct experiments with their customers as a powerful new means of value co-creation. This ability to rapidly test ideas fundamentally changes a company's mindset and approach to innovation. Rather than agonize for months over a choice, or model hypothetical scenarios, a company can simply ask its customers and get answers in real time, significantly compressing the "hypothesis-to-experiment" cycle time.

Open access to Big Data is critical. Companies will increasingly need to integrate information from multiple data sources, often from third parties including competitors, and the incentives have to be in place to enable this. This is a continuation of the open innovation meme—call this Open Analytics—let competitors share the same information and then compete on their interpretation of it.

The intersection of predictive analytics and cloud computing forms a compelling value proposition. The initial effort on the part of the business analyst is modest in comparison with provisioning an in-house predictive analytics platform. First, the march toward wider deployment of predictive analytics is enabled by deployment in the Cloud for much lower costs than in-house approaches. Second, the march toward engaging and solving sophisticated predictive problems is enabled by innovations in implementing advanced algorithms across much larger data sets supportable by Cloud processing capabilities.

Beware. Discovering and determining meaning is a business task not a statistical one. There are no spurious relationships between variables, only spurious interpretations. How to guard against misinterpretation? This requires deep experience and knowledge of human behavior (customer), the business (service), and the market (intersection of the two). The more challenging (competitive, immature or commodity-like) the business environment, the more worthwhile are the advantages attainable through predictive analytics. The more challenging the business environment, the more important become planning, prototyping, and perseverance. The more challenging the business environment, the more valuable becomes management perspective and depth of understanding. And all of this is magnified including value when openly, transparently shared in a collaborative network.

The latest advance in predictive analytics comes in the form of genetic algorithms deployed to address issues in business analytics. Genetic algorithms mimic the parallelism of variation and selection characteristics of biological evolution. They have been used in skunk works on Wall Street to model the behavior of individual securities. This is yet another example of the evolution from command-and-control models (think statistics) to collaborative emergent models (genetic algorithms) in identifying opportunities for innovation.

In addition to genetic algorithms, other forms of "artificial intelligence" will be needed to make sense and take action in the world of Big Data: Personal software agents acting on our behalf, autonomous intelligent objects, multi-agent systems, ontologies for semantic mediation and Bayesian inference techniques.

An architecture appropriate for Big Data and the Internet of Things goes well beyond what we usually associate with IT architecture. Let's glimpse at the needed changes cited in *Architecting the Internet of Things*, "The Internet has changed our business and private lives in the past years and continues to do so. The Web 2.0, social networks and mobile Internet access are just some of the current developments in this context. Ubiquitous computing and ambient intelligence have been fields of research where changes of computing in everyday situations have been examined. Today, the Internet of Things is a foundation for connecting things, sensors, actuators, and other smart technologies, thus enabling person-to-object and object-to-object communications.

"Will the Internet of Things make our lives easier? Or will it just be another component in a world of information overflow? Currently, the Internet of Things is all about information visibility—it is not [yet] about autonomous decision-making.

"To relieve us from everyday decision tasks and to avoid delays between information availability and decisions, new methods and technologies need to be integrated. For example, in logistics, *autonomous cooperating logistic processes* use decentralized and hierarchical planning and control methods. The combination of autonomous control and the Internet of Things can provide a higher level of infrastructural robustness, scalability and agility."

tinyurl.com/6tyj26p

The stuff of autonomous logistics objects and other complex environments is simply overwhelming if left to mere humans alone. We'll need what former IBMer, James Martin, dubbed as "Alien Intelligence" to cope, and the various fields of intelligent agent technology have been "solutions looking for problems" for years.

Now their time has come.

Oh, by the way, where are we going to store all this stuff? It seems that breakthroughs in storage technology are needed. But help is on the way. With a discovery that could some day fundamentally alter the scale of mass data storage, nanotechnology researchers at IBM say they have found a way to store a bit of information in as little as 12 magnetic atoms (atomic data storage).

That's a radical improvement over today's storage devices which, IBM argues, require about a million atoms to hold a bit of information. For those keeping score at home, IBM's discovery could mean storage could one day be possible at 1/83,000th the scale of today's disk drives.

But there's more than just quantum leaps needed in storing Big Data, there's also the challenge of *distributing* it. Today the wireless Internet depends on the *radio frequency spectrum*, a scarce-resource spectrum coveted by the telecom companies that also charge what they please for access, lobbying that there just isn't enough bandwidth and that they want to control it as a limited resource (think $). Hogwash, says Edinburgh University professor, Harald Haas.

Haas wants to deliver broadband *wireless data from every light bulb*!

Forget the radio spectrum. We are heading to a saturation point in terms of how efficiently we can use the radio frequency spectrum. It is forecast that by the year 2015, we will transmit six exabytes—six billion, billion bytes—every month through wireless networks. This is a ten-fold increase on the amount of data we send now.

In order to meet this increased demand, we need either 10 times more radio frequency spectrum for commercial wireless networks, or we have to make the existing radio frequency spectrum

10 times more efficient. The first is impossible—most of the available radio frequency spectrum is already used. The second option is difficult to achieve, as existing wireless technology is very sophisticated, and it has been shown that further improvements are often offset by unmanageable complexity. The only way out of this is to find new ways to transmit data wirelessly. Fortunately, the electromagnetic spectrum not only incorporates the radio frequency spectrum, but also includes the *visible light spectrum*, the best known transmitter of which is the sun!

Haas says, "It should be so cheap that it's everywhere. Using the visible light spectrum, which comes for free, you can piggy-back existing wireless services on the back of lighting equipment. If all the world's incandescent light bulbs were replaced by LED, the energy saved would be equivalent to that produced by more than 100 nuclear power stations. However, this is not the only advantage of LEDs. These lights are semiconductor devices similar to transistors, which are commonly found in devices such as TVs, laptops or smart phones. Like transistors, LEDs can be switched on and off very quickly.

tinyurl.com/3rcf5z5

"We have harnessed this feature to develop novel techniques that enable ordinary LED light bulbs to wirelessly transmit data at speeds many times faster than WiFi routers. We have named the new technology Li-Fi (light fidelity). In our lab, under ambient light conditions, we are able to achieve data speeds of 130 megabits per second. If all light bulbs were able to do this, it would create a simple, energy-efficient solution to the lack of available radio frequency spectrum for future wireless broadband communication. The new Li-Fi technology utilizes existing infrastructures and as a result the installation costs are minimal, let alone the reduced cost of the technology as it does not require an antenna.

Caveat telcos!

In addition to meeting the exponential demand for broadband access, another challenge must be met in order to overcome the Balkanization of the Internet by the Internet providers themselves who seem to be in the process of creating gated communities under the guise of "tailoring or personalization." Facebook does it. Google does it. Amazon does it. Yahoo news does it. If one person searches for something, and you search for that same something, even right now at the very same time, you'll get very different search results. Even if you're logged out, one engineer said, there are 57 signals that Google looks at—everything from what kind of computer you're on to what kind of browser you're using to where you're located—that it uses to personally tailor your query results. Think about it for a second; there is no standard Google anymore. And you know, the funny thing about this is that it's hard to see. You can't see how different your search results are from anyone else's. And this moves us very quickly toward a world in which the Internet is showing us what it *thinks we want to see*, but not necessarily what we *need to see*. As Eric Schmidt said, "It will be very hard for people to watch or consume something that has not in some sense been tailored for them."

This *filtering*, more than anything else, may be the counterpoint to the Cloud being a source of innovation. It may in fact impede innovation by filtering out the many weak signals that point to new needs and opportunities. Companies that want to innovate must seriously take into account the impact of Internet filtering.

Back in ancient, pre-Internet times, broadcast and print media *editors* served as the "gatekeepers" that controlled the flow of information. That was a very important role that earned news reporter Walter Cronkite the label of "the most trusted man in America." Today's gatekeepers are filtering algorithms and the most important criterion is "relevance." What we're seeing is more of a passing of the torch from human gatekeepers to algorithmic ones. And the thing is that the algorithms don't yet have the kind of embedded ethics that the editors did. So if algorithms are going to curate the world for us, if they're going to decide what we get to see and what we don't get to see, then we need to make sure that they're not just keyed to "relevance." We need to make sure that they also show us things that are *uncomfortable* or *challenging* or *important*; they need to present *other points of view*. We need the new gatekeepers to encode that kind of responsibility into the filtering code that they're writing.

We really need to make sure that the new gatekeepers, be they our sources of information or our own analytical processes, these filtering algorithms, have encoded in them a sense of the public life, a sense of civic responsibility and ethics. We need to make sure that they're transparent enough that we can see what the rules are that determine what gets through our filters. And we need those filters to give us some control so that we can decide what gets through and what doesn't. This is because we really need the Internet to be that thing that we all dreamed of it being. We need it to connect us all together. We need it to introduce us to new ideas and new people and different perspectives. And it's not going to do that if it leaves us all isolated in a Web of one. Watch Eli Pariser, author of *The Filter Bubble,* take on these critical issues that could very well trend the future of democracy itself in the new world of Big Data.

tinyurl.com/4xkj8hm

Takeaway

Way back in 1991, Yale professor and unabomber victim, David Gelernter, published a book called *Mirror Worlds: Or the Day Software Puts the Universe in a Shoebox* where he wrote, *"Someday soon you will look into a computer screen and see reality. Some part of your world—the town you live in, the company you work for, your school system, the city hospital—will hang there in a sharp color image, abstract but recognizable, moving subtly in a thousand places. This Mirror World you are looking at is fed by a steady rush of new data pouring in through cables. It is infiltrated by your own software creatures, doing your business. ...Mirror worlds will transform the meaning of 'computer'."*

One could argue that a big data, *quantitative input* driver to innovation is a Google model. It relies on rapid experimentation and data. Google constantly refines its search, advertising marketplace, e-mail and other services, depending on how people use its online offerings. It takes a bottom-up approach: customers are participants, essentially becoming partners in product design either directly or in the patterns of usage they create.

Google speaks to the power of data-driven decision-making, online experimentation and networked communication and collaboration. Thomas R. Eisenmann, a professor at the Harvard Business School calls this "business and management innovations lubricated by technology."

You can think of it as an engineering driven model, based upon convergent thinking. It is a very analytical and rational process. It is quantitatively driven and generally sequential in execution. Design is primarily focused on constraints (real or perceived limits as to what can be done) and is very objective – focused on what is known. This is evident in the amount of specific details generally involved in this mode of innovative thinking.

Qualitative Input. Equally important to innovation are the *qualitative*, inspirational "Ah ha" moments. Compared to the Google model, this could be called the Apple model. The Apple model is more edited, intuitive and top-down. Like Henry Ford's comment that his customers thought they wanted a faster horse, Steve Jobs was noted to say, "It's not the consumers' job to know what they want."

This model employs more divergent thinking – it is both creative and intuitive (which may be the wonderful analog computer in our heads performing the big data analytics function of quantitative input – only in a different manner). This approach to both idea generation and evaluation is qualitative and subjective ("It feels right", "It's cool"). While quantitative engineering processes and decision making tends focus on the constraints that have to be lived within, divergent, qualitative approaches rely upon "conceiving the possibilities."

While a Google model tends to convey very specific and focused outcomes, an Apple model is very holistic (much of the success of the Mac and the "iXXXs" can be traced back to the tight integration of functionality, hardware, software and now with iCloud (and its earlier manifestations like "Me") services. Initial innovations at Apple are generally driven by conceptual abstractions ("wouldn't it be neat") rather than specific problem solving.

One issue with a pure Google model is that the closest connections and deepest interactions with customers and partners usually occur at the edges of the organization in exceptional (read small numbers of) interactions which do not show up in the "big data" clearly. These people at the edges hear the "voice of the customer" every day and they actually experience the inefficiencies or inadequacies of processes or systems. Also, the newest and youngest employees often begin their careers in customer service, sales or field service positions. In their early tenure, these people are unimpeded by "the way we've always done it" and are unaware of the political obstacles of the organization. However, their knowledge is experiential (read qualitative) rather than analytic (read quantitative). In fact, "It's a lot of data crunched in a nonlinear way in the right brain," as described by Erik Brynjolfsson, director of the M.I.T. Center for Digital Business.

A second issue with the pure engineering model for innovation is that the innovators tend to focus on top-line opportunities such as developing new products, services or technologies, because that is what is necessary to generate an ROI or ROE (to get funding). Therefore, the enterprises that address innovation only through an analytic approach may miss many opportunities at early stages.

Quantitative approaches to innovation also tend to get lost in the numbers, and the management "to the numbers" become the driving force in the organization instead of what is really trying to be accomplished. To quote Goodhart's law once again, "When a measure becomes a target, it ceases to be a good measure." With innovation, companies tend to loose what they need to do brilliantly (problems customers don't even know they have yet) from what they just need to be good enough at (problems with multiple suppliers of solutions) to succeed and create great value for their customers.

Watch Silicon Valley marketing consultant, Geoffrey Moore at the Business of Software 2009 conference that addresses that very subject:

tinyurl.com/yerkjsm

14. The End of Management —As We Know It

"While 'modern' management is one of humankind's most important inventions, it is now a mature technology that must be reinvented for a new age."—The Management Innovation eXchange *tinyurl.com/2dp6mc8*

Technology is the making, usage and knowledge of tools, techniques, crafts, systems or methods of organization in order to solve a problem or serve some purpose. According to this definition, the alphabet is a technology that certainly has served a purpose for humankind.

The invention of the steam engine gave rise to factories and mass production, which in turn gave rise to a command-and-control style of management.

Hmm? Can we thus think of management as a technology?

Mass production was based on economist Adam Smith's notion of specializations and efficiency, where units of work were controlled from the top down. Just as a management structure and style was adapted to the factory as the focal point of economic activity, management must adapt to the Cloud as the focal point of economic activity. An emergent organizational form, the "Cloud Enterprise," will have changed its management structures and styles to become organic networks rather than hierarchical, function-divided monoliths.

The problem we are facing today is that the organizations that generated all these mass-produced products were not designed for the current velocity of business.

They were not designed to listen, adapt and respond. They were designed to create a ceaseless, one-way flow of material goods and information. Everything about them has been optimized for this uni-directional process, and product-oriented habits are so deeply embedded in our organizations that it will be difficult to root them out.

Today, thanks to the Internet, the kind of creativity and innovation that used to take place primarily within corporate walls, increasingly takes place over large amorphous networks of peers.

Millions of people already join forces in self-organized collaborations such as Linux and Wikipedia that produce dynamic new goods and services that rival those of the world's largest and best-financed enterprises. And if the masses can peer-produce an operating system, an encyclopedia, a mutual fund, and even physical things like a motorcycle, one should carefully consider what might come next.

You could argue that we're becoming an economy unto ourselves—a vast global network of specialized producers that swap and exchange services for entertainment, sustenance, and learning.

The lesson for business leaders is that the old monolithic multinational that creates value in a closed hierarchical fashion is dead.

Winning companies today have open and porous boundaries and compete by reaching outside their walls to harness external knowledge, resources, and capabilities.

Rather than do everything internally, these companies set a context for innovation and then invite their customers, partners, and other third parties to co-create their products and services.

What does this mean for Management?

In what we may call the Business Singularity or the Business Network to keep it from sounding too futuristic, the Cloud makes it possible for multiple companies to come together to work as *one* value delivery system, not just for efficiency, but more importantly for responsiveness and innovation.

But such new organizational forms cannot be managed like the factory of old, for each participating business runs on its own clock, using its own internal rules and methods.

> Industrial Age command-and-control leadership gives way to connect-and-collaborate, where every member of a business team is a "leader."

In the Cloud, leaders don't give commands, they transmit information, trusting the team members' competencies and gaining accountability through the *invisible hand of transparency*. Let us not forget that it was Adam Smith who introduced us to the "Invisible Hand of Markets" one of the earliest expressions ever recorded of emergent behavior and decision making by a network of independently acting agents in markets.

True leadership is about cooperation, not control. In this chapter we'll explore the end of traditional management and the rise of self-organized and adaptive Bioteams that are based on nature's best designs. No, this isn't science fiction; it's the future of organizations getting stuff done. It's about unleashing human potential in our organizations. It's also about passion.

Not all innovations are created by advances in technology. For example the biggest business innovation in the last 100 years wasn't a new gadget—it was "management." Sometimes inventions that come in the form of ideas are so significant that we actually forget they were *invented* at one point in time. And if an invention affects our lives by having economic value it becomes a business innovation.

Years ago, thanks to Alfred Sloan, Winslow Taylor and Peter Drucker, almost all the tools of modern management were invented. Pay for performance, capital budgeting, task design, divisionalization, brand management and all the methods and tools that we find here today in organizations, all of those were invented before 1920. And most of them demanded advances in technology if the ideas were to be implemented in the real world. Then those advances in technology opened up new possibilities for further advances in management techniques.

London Business School professor, Gary Hamel observed, "Go back to the fundamental question—what problem has management invented to solve, 100 years ago? What problem were those people trying to solve? I can tell you—it wasn't the problem of being an adaptable, innovative, inspiring place to work.

"The problem they were trying to solve was—how do you turn the human beings into semi-programmable robots. How do you make the farm hands and house maids and the crafts people to show up on time and tend to machines, to do the same thing over and over again? That's how we succeeded. So, that's the DNA of today's organization—that's the goal you're supposed to serve.

"Now, you have to go back and challenge some of those fundamental principles:

No. 1 you have to have an aspiration.

No. 2 you have to be a contrarian.

No. 3 you have to learn from the fringe."

"Innovation in management, like every other sort of innovation, starts at the fringe. It doesn't start at the mainstream. It will happen 2 or 3 levels down but it's not going to be coming from the top of the organization. Whether it's art, music, literature or fashion – the future happens on the fringe. If you want to really see the future of management, you have to look at some of the really unusual places. The best place is the Web. I mean, management is a feudalistic system. If you think about the Web—it's not so much."

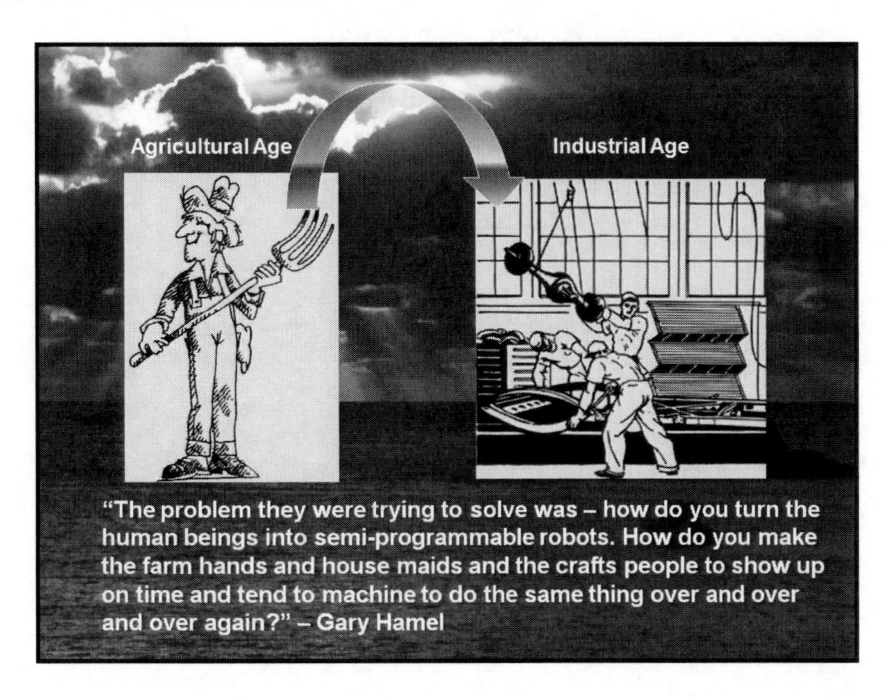

Fortune Brainstorm Conference
Michael Dell and Prof. Gary Hamel

Watch: Prof. Hamel

tiny.cc/2ygph

"The Web is the global operating system for innovation. So, I have a feeling that the kind of values that characterize the Web we have to bring those into our organizations. The values of openness, meritocracy, flexibility, collaboration—all those deep values—forget about technology—all these deep values have to become values of our organizations. Why? Because the Web is already adaptable, already innovative, it's already amazingly engaging. It has all of the characteristics of the organizations of the future." — Prof. Gary Hamel

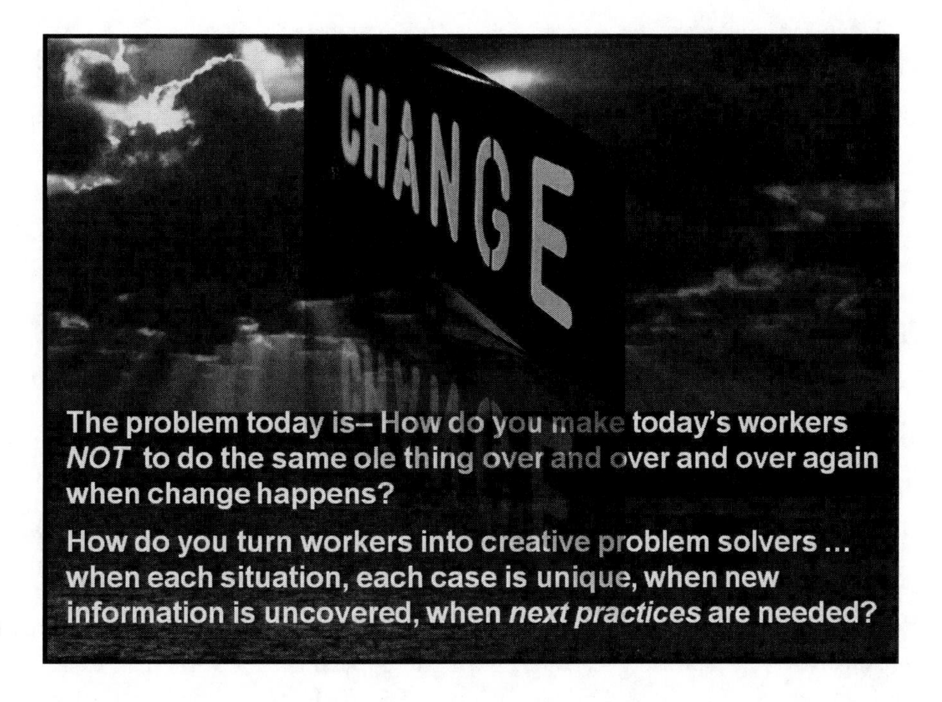

So, if you want to be a master innovator, the question to ask is, "How do you take—not just the tools because they are sometimes kind of superficial—but how do you take these deep principles on the Web and bring them into our organizations?"

Open Leadership.

Transparency, authenticity, openness. According to former Forrester analyst, Charlene Li, management innovation, "... is about giving up control yet staying in command... it's about inspiring not controlling." In an interview with professor Gary Hamel, Li summarized her views on the transformation from command-and-control management, "There's always been this tug of war in leadership between having power centered in one person versus the power of responsibility going out to the people in the field who actually get the work done. So how do you go back and forth between this command-and-control kind of philosophy, especially for first-time leaders? They really come in and say, 'I'm going to micromanage everything, command everything.' Yet, more mature leaders, the people who have experience, realize very quickly that 'I have to let go of control.' In fact, I don't really control anything. I lead because I have credibility, because I have faith, I have relationships. So what if you started leadership from that premise, that is not about control, but giving it up, being more open about the information you share and how you make decisions?

"I like to say you're not in control, but how do you get back in command? Command is about getting things done. It's not about controlling things. It's having that relationship with people so that when you ask for something to be done, they're inspired, they follow you, they trust you, that you are given the right decisions. And that's traditionally where leadership comes from, that right to be able to command, it's a right.

"Command says that, when you say charge up that hill, people will actually do it. In the end, that soldier has to make those legs move in the face of tremendous risk and fear. And the only way they will do that is they believe in you. They really, truly believe in you. And some soldiers will say, I'm less afraid of charging up that hill than of getting my head chopped off by the sergeant. But the best soldiers will be the ones who believe that you are leading them in the right direction.

"In the end, who are you going to follow? Somebody who you are really inspired by, who you trust and you have that relationship and the ones that you want to spend the time and investment, the personal investment going above and beyond. Is that the kind of relationship you want with the employees? I think so.

"There's always going to be room for the hierarchy because work needs to get done in a fairly systematic way. But there's also room for a new type of leader, and I'll give you a very poignant example.

"Salesforce.com has a product called Chatter. It's sort of an internal way of using Twitter. So when Marc Benioff had his strategic offsite with the top 300 leaders of the organization, he invited the top 25 Chatter users, based on their followers and their usage. And so he had those people sprinkled in, and they were a part of the leadership.

"And I asked him afterwards, where do you take this? If your chatterati and your SVPs are sitting here creating the same amount of value to your organization, where do you take it? He said, we may have to rethink the way that we think about leadership, the definition of it. We may have to pay the chatterati as much as the SVPs because they have as much value.

"And I think it goes back to, if you get things done not because of authority but because of influence, then you've got to pay the people who have influence. And you develop your learning and development systems based on influence, in addition to authority. I mean, authority has to still be there because there needs to be some sort of chain of command to help things. I mean, companies like [W.L.] Gore, they don't have any hierarchies, right? It's completely done by consensus. And yet they naturally come in, based on the natural work that needs to get done. So it's work-centered hierarchies that naturally fall out based on who is doing it. But it can happen much faster now because of these [social] technologies.

"And the beautiful thing is can you marry the two together of having hierarchies that are there to get work done in a very systematic way, but also marry them across divisions and silos through these natural connectors."

tinyurl.com/6vgledf

It seems that the universe continues to evolve under certain immutable laws. These laws apply to everything: biology, physics, and social systems that, in turn, include these things we call an enterprise and business process management.

An overlay of end-to-end process management onto existing functional organizations has its rough edges, to say the least. In fact, the transformation to a process-managed enterprise could really mean the End of Management, as we know it.

What might a process-managed org chart look like? If you consider that today's average value chains consist of over 20 autonomous companies, each running according to its own clock, then we'll have to unbundle our traditional hierarchal models that structure our organizations, where information flows through filters up and down the command chain, and look to nature for new patterns that have been under development for millions of years. By mimicking the designs of nature, the new org chart will no doubt look far more like the complex adaptive system made up of autonomous, self-organizing, self-managed teams found in nature.

While operational innovation, via business process management, is the cornerstone of 21st century competitive advantage, operational innovation will be for naught without innovation of another kind, Management Innovation. When responsiveness trumps efficiency, hierarchical command-and-control management systems fall flat. In a world of hyper change, centrally controlled hierarchies simply cannot see the opportunities or move quickly enough (remember the Soviet Union). Even if they utilize advanced business process management (BPM) systems, their management processes and structures cannot adapt with the required speed.

What pioneering companies are doing is to replace top-down pyramids with trellises of latticework, where each lattice represents an autonomous, self-managed team.

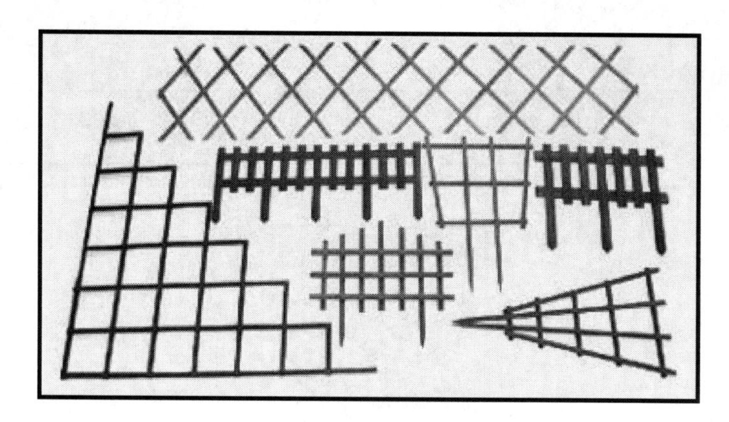

Sound farfetched and anarchistic without central control? You bet. But, guess what? That's a mirror image of the real world, a world made up of complex adaptive systems.

Turning to the Santa Fe Institute's John H. Holland: "A Complex Adaptive System (CAS) is a dynamic network of many agents (which may represent cells, species, individuals, teams, firms, nations) acting in parallel, constantly acting and reacting to what the other agents are doing. The control of a CAS tends to be highly dispersed and decentralized. If there is to be any coherent behavior in the system, it has to arise from competition and cooperation among the agents themselves. The overall behavior of the system is the result of a huge number of decisions made every moment by many individual agents."[1]

Complex adaptive systems are managed without managers!

Can businesses manage without managers? Indeed. Just look at W. L. Gore, the company that makes Gore-tex, the fabrics that keep you warm and dry, but that also breathe, and Glide dental floss. In 1958, former Dupont engineer and geek (e.g., no MBA), Bill Gore would forego a ladder-like hierarchy and create an organization with a flat, lattice-like organizational structure where:

- There are no "employees," and everyone shares the same title: associate.
- There are over 8,500 associates that have turned out over a thousand innovative products.
- There are neither chains of command (pyramids or reporting structures) nor predetermined channels of communication. Anyone can talk to anyone, anytime.
- There are a large number of small, autonomous, self-organizing teams that function as a web of startup companies.

- There are no bosses, no V.P.s, executives, or managers—there are "sponsors" and "leaders." Associates are only responsible to their teams: Everyone's the boss, and no one's the boss. There are no slackers.
- There are no standard job descriptions or "assignments." Associates make sets of "commitments" to their teams, and sometimes unorthodox titles emerge, e.g., a "category champion."
- Associates choose to follow leaders rather than have bosses assigned to them.
- Performance reviews are based on a peer-level rating system.

What's really going on that makes a real difference at Gore? The company has closed the "information gap." Humans are unique in their ability to manipulate information outside the body, and are constantly hungry for information and narratives. While we speak of a growing wealth gap in today's economies, the volume and depth of information available to senior executives compared to the shop floor worker is staggering in typical corporations. Left starved for information, it's little wonder why curiosity and creativity remain in lock-down among the rank and file in most corporations. As John Caddell writes in his *Shoptalk* blog, "Perhaps it's concern for confidential information leakage, or for PR fallout, or that management simply doesn't trust in the employees' [Theory X] ability to add value to innovation."[2]

Lots of companies have posters in the break room with slogans such as Empowerment, Initiative and Teamwork. At Whole Foods, those aren't just your typical empty slogans, they are the foundations of its operations.

The company has a "Declaration of Interdependence" that states its commitment to diversity, community, and saving the planet. Its salary caps limit executive compensation to 14 times the average pay across the company. But it's no airy-fairy socialist experiment in capitalism. Whole Foods is a hardball competitor, both inside and out. It harnesses the passion, energy and competitive spirit of *all* of its people, while typical companies bury those factors under a pyramid of management hierarchies and bureaucracies.

Whole Foods secret sauce?

It's all based on autonomous, decentralized teams. The team, not the typical management hierarchy, is the building block of the entire company. Each store is its own profit center made up of self-managed teams that include seafood, produce, grocery, prepared foods, and so on. Each team selects its own leaders and its employees. Team leaders, in turn, make up the store's team; store leaders make up regional teams; and regional team leaders make up national teams.

Such distributed power and authority would result in chaos in traditional organizations, but *transparency* is the invisible hand of accountability and management control. Financial information, including store and team sales, profit margins, and, believe it or not, salaries, are available for all to see. It's like seeing batting averages in major league baseball. Such scorecards lead to intense competition as teams, stores, and regions compete vigorously to outdo each other in quality, service, profitability. But it's more than just friendly competition and community spirit, scorecards translate into bonuses, and promotions. That's democratic capitalism in action. Peer pressure trumps bureaucracy, combining democracy with competitive discipline.

In each store, the power is concentrated in the hands of teams. Teams approve new hires, who require a two-thirds "yes" vote after a month long trial period. And teams are rigorous about their member's selection because financial incentives are tied to team, not individual, performance. And, to build trust, every member of every team knows the salary of everybody in the team, and in the entire company.

HCL Technologies is one of the fastest growing IT services company having 77,000 employees working in 26 countries generating revenues of $3.5 billion. Turning the organizational pyramid upside down, HCLT's management innovation rests on the notion of *reverse accountability*. At HCLT all employees rate their bosses and their bosses' bosses—and all those ratings are published online.

Also, there is an interesting ticketing system in this company. If first-level employees disagree with the decision of their bosses or feel they have been treated unfairly or it's taking too long to process their expense claims any employee can fill out a ticket. The tickets are visible, transparent to the whole organization, across the globe. And a given ticket can only be closed by the employee that raised the ticket. Any ticket that doesn't get closed in 24 hours is escalated to the next level of management.

That's reverse accountability: people hold their managers accountable, helping them succeed in their jobs.

Many companies have implemented the proverbial "Suggestion Box," where anyone can anonymously submit suggestions. But the classical suggestion box is little more than a fancy trash can as who knows what's been suggested, much less what's been acted on.

HCLT, on the other hand, has a transparent suggestion box in the Cloud where all is visible.

HCLT believes that all of the value is created at the interface of the employee and the customer, and so management's job is to encourage the innovation there. They explicitly say to their employees, "You are more important than managers." Thus the mantra of this Co. HCL Technologies is "Employees first, customer's second." How often does somebody say that? The CEO Vineet Nayar stood in front of this customer group, the CIO's of the big companies around the world and told them, "I'm sorry, but for me you don't come first."

Nayar recalls a visit he once made to a friend's home in Amsterdam. It had huge windows that looked out on the city's main canals. The home was flooded with light, and the rooms felt very exposed to people outside.

"Why do you have such large windows?" he asked.

"It keeps the house clean," the friend said.

Nayar took his answer to mean that the bigger your windows, the more glass you have in your house, the more visible the dirt will be—to you and to everyone who visits or passes by. If you can see the dirt, you will be much more likely to get rid of it.

A transparent house has a dramatic effect on the culture inside. And that's exactly what HCLT did—threw the window of transparency open and let the light in.

While command-and-control might work well in a coal mining operation where employees are sequestered underground, connect-and-collaborate applies where employees are the "product" that embodies value.

Got Team? According to the distinguished Indiana University technology professor, Dr. Curtis Bonk, "This is the age of employee participation, multiple leaders and yet no leader, and prompt communication, as well as the technologies that make all this possible." Got team? You'd better. To succeed in today's dynamic, technology-enabled environment, you must be able to function in and through teams. But, if we stick with our current pyramid-style designs of our organizations, we will not be able to meet the growing needs of our communities in the high-change global economy.

The discipline of *bioteaming* offers a vision of what successful teaming experiences look like in the interconnected world of the 21st century. A February 2008 *Business Week* feature, "Using Nature as a Design Guide," focused on how the "biomimicry" design movement helps companies look to the natural world to help make their businesses green. The feature reported on the pioneering work of Janine Benyus, biologist-cum-evangelist, the driving force behind the movement. In her writings she detailed how companies could study nonpolluting, energy-efficient manufacturing technologies that have evolved in the natural world over billions of years.

Now enter Ken Thompson, the former CIO with Reuters, who over the past ten years has taken the field of biomimicry from innovation related to physical things on to the realm of social structures. Thompson takes ideas from Nature about how groups perform and operate, and applies them to enhance how humans can work together in groups and teams.

Thompson's "bioteaming" is about designing and implementing organizational teams that operate on the basis of the communication principles that underpin nature's most successful groups. Spot the common theme: the waggle dance of honeybees, the pheromone trails of ants, the one-way information bursts of migrating geese.

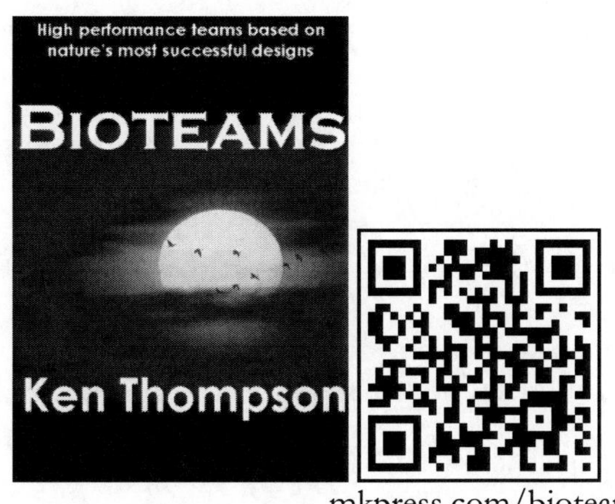

mkpress.com/bioteams

According to Thompson, nature's teams have characteristics that are not usually present in organizational teams: *Collective Leadership:* Any group member can take the lead. Nature's groups are never led exclusively by one member. Instead different group members lead as needed. When geese migrate it is well known that the goose leading the V formation rotates. However, this is not just because they get tired and need to fly in another goose's slipstream for a while. The real reason is that no one goose knows the whole migration route. Collectively, between them, they know the migration route but no one individual knows. So a goose leads the part of the journey where it knows the way and when it recognizes "I don't know where to go next" it flies back into the V and waits for another goose to take over.

This is "Collective Leadership," the right leader for the right task at the right time.

Takeaway

The human species seems to be the only species that trusts in a single leader (or small management team) to know the whole path, on behalf of the community. Multi-Leader groups possess much greater agility, initiative and resilience than groups that are only led by a single exclusive leader. In multi-company value chains, collective leadership takes place in the Cloud.

References

[1] *Complexity: The Emerging Science at the Edge of Order and Chaos* by M. Mitchell Waldrop.

[2] http://shoptalkmarketing.blogspot.com/2007/10/on-gary-hamels-future-of-management_19.html

15. Services Innovation

Forward-looking companies are turning their focus to the emerging field of Services Sciences, an interdisciplinary approach to the study, design, and implementation of services systems made up of people and technology to create and deliver value.

Irving Wladawsky-Berger, Chairman Emeritus of the IBM Academy of Technology, prefers to think of cloud computing as offering all kinds of *Services-as-a-Service*: consumer services, business services, government services, health-care services, and so on. As advanced societies have transitioned from agricultural to manufacturing economies, the current transformation is to a service economy.

The services sector makes up 70 to 80% of GDP in advanced economies. The best way to define the services sector is to understand what it is *not*. It's not agriculture or manufacturing or construction, the *shrinking* sectors.

In the U.S., agriculture accounts for only 1.4% of the gross domestic product and less than 2% of employment. Meanwhile, 4.3% of firms fall into the manufacturing sector, accounting for 12.5% of employment. In the UK, agriculture accounts for around 1.5% of employment, manufacturing for around 10%, and services for over 80%.

In short, up to about 75 % of wealth in industrialized countries is created *not* by growing food or making things, but by performing services: teaching, designing, delivering health care, banking, retailing, consulting, delivering IT services and so on.

The link between science and agriculture and manufacturing is now a given and was ushered in over the last 200 years with the advent of the Industrial Age. But what about the link between science and services? The information technology to drive innovation in services have really only been around since the Web came on the scene in the mid-1990s.

IBM has played a leading role in this emerging field, which is no surprise for IBM had to transform itself as it looked over the abyss in the 1990s. Lou Gerstner left RJR Nabisco to become IBM's CEO in 1993 and led the way to unlocking talent inside the once stodgy company. Hmm, a move from the "cookies and crackers company" to mighty Big Blue. What was up with that? *Services.*

Today, IBM pulls in the bulk of its revenues from services. Noting the lessons learned by IBM, HP bought EDS in 2008 and Dell bought Perot Systems in 2009. And in June 2009, Howard Smith, coauthor of *Business Process Management: The Third Wave* was invited to speak at GE Global Research's *Whitney Symposium09* in Niskayuna, New York, a 600 acre research campus on the banks of the Mohawk river... the legacy of one Thomas Edison. It was an eye-opening experience (1,000 PhDs representing 22 disciplines in one campus). And you guessed it, the theme for the Symposium was *The Engineering of Customer Services.* GE gets it.

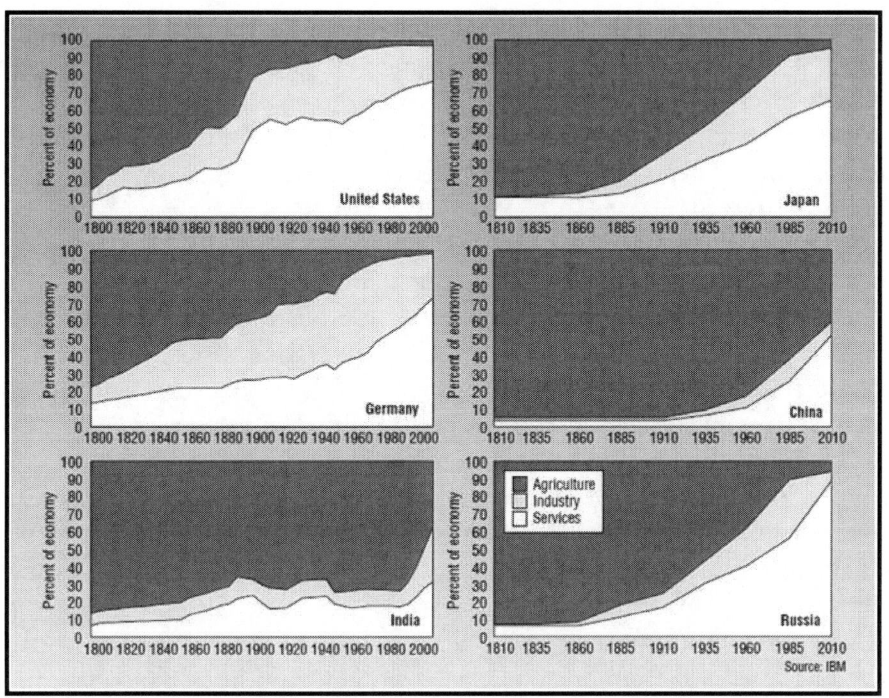

Services Represent a Growing Segment of the World's Economies

Peter Fingar's keynote talk, *Business Process Management and Systems Thinking* was followed by Robert Morris, V.P. of Services Research and former director of IBM's legendary T. J. Watson Research Lab. Morris' talk was *The Transformation of IBM to a Technical Services Company*. To say the least it was an intense two days on the banks of the Mohawk river.

Why, you might ask, were "BPM experts" at a *services* symposium? The following call-out from an IBM publication says it all in one sound bite:

> "Service science melds technology
> with an understanding of business processes."

While much activity in BPM centers on supply chains, logistics and other physical means of optimizing value delivery systems, that's not where the innovation action is in a service economy.

Yet, we have done little to take on services as the object of BPM initiatives as we don't really know much about what it takes to bring innovation to service processes. Most companies today are designed to produce high volumes of consistent, standard outputs, with great efficiency and at low cost. Even many of today's services industries still operate in an industrial fashion. Schools efficiently produce standardized students. Hospitals efficiently move the sick and injured through a diagnostic-and-prescriptive production line. Drive-through restaurants move drivers quickly and efficiently through an order-fulfillment pipeline. But most of these examples are not really *service processes* at all. They are factory-style processes that treat people as if they were products, moving through a production line. Just think of the last time you called a company's "customer service line" and ask yourself if you felt well-served.

Sure, many services require some level of production efficiency, but service processes are not what we most often think of as pre-defined business processes. They are *experiences.*

Unlike products, services are often designed or modified as they are delivered; they are co-created with customers; and service providers must often respond in real time to customer desires and preferences. Services are contextual—where, when and how they are delivered can make a big difference. They may require specialized knowledge or skills. The value of a service comes through the interactions: it's not the end product that matters, so much as the experience.

As explained in the book, *Business Process Management: The Third Wave*, "A business process is the complete and dynamically coordinated set of *collaborative* and transactional activities that deliver value to customers." Well it's the *collaborative* part that is the challenge when it comes to services. Just consider the very nature of service processes:

- *Intangibility:* May be some combination of both intangible and tangible results or processes.
- *Heterogeneity:* Outcomes vary from one knowledge worker to another. Skills-based routing is vital.
- *Simultaneity:* Production and consumption occur at the same time. Complex Adaptive Systems: self-production, self-organization. Perceived control. Decision-making processes of the customer.
- *Perishability:* May be consumed immediately. Can't be stored. Once the event or time has passed, the opportunity is gone forever.
- *Customization:* Services are almost always customized. A transaction equates to a new "product."
- *Labor Intensive:* Products are capital intensive, while services are labor intensive—People doing things for other people.
- *Resource Pull:* Appropriate services must be pulled on demand.

In physical supply chains customer demand is rarely perfectly stable, so companies do their best to make-to-forecast. Unfortunately, forecasts are rarely accurate and thus companies compensate by having a "safety stock." Going from the ultimate customer down to suppliers and suppliers' suppliers, each supply chain participant sees greater variation in forecast demand and thus has greater need for safety stock. In other words forecast variations are *amplified* as one moves downstream in the supply chain – the *bullwhip effect.*

While modern technology can allow companies in a supply chain to move to a *make-to-demand*, versus *make-to-forecast* model, how can the bullwhip effect be tamed when it comes to meeting the demand for immediate, perishable services? The *bullwhip effect* is even more accentuated in services industries where processes are less likely to be defined and a single service may require multiple unique processes.

What's the equivalent of *safety stock* in a services chain? Because the knowledge workers cannot be *stockpiled*, that's where the emerging field of service science comes in to help answer questions of: "What types of technologies can come to bear on service processes?" Let's take a look at what's involved in the service value chain.

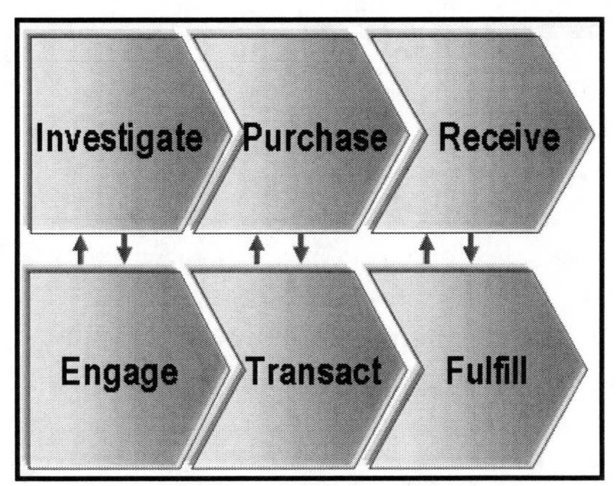

The Service Value Chain

> A service process is a series of unique states
> involving the *co-production* of the provider and the consumer.
> Each "transaction" is a new product in itself,
> and the customer is a co-producer of the value rendered.

Because services must address increasingly demanding customer preferences, companies must find ways to make them more granular, as well as easier to bundle with other services. Customers want services to be convenient for them, not for you, the service provider. Too often, when we think of a service process, we think of what the provider must do, but such thinking results in frustration for the customer – who will defect in an instant. When we think of service processes, think of the customer as a co-producer of value. Think of migrating from transaction chains to information chains, and then on to knowledge chains (or peer-to-peer knowledge webs to be more precise). Moreover, think about the Cloud and cloud computing technologies. The automation of business processes is a key enabler of the Cloud phenomena.

Without process, the Cloud remains a passive environment
that undoubtedly saves money and removes some of the operational headaches, but does little else.
The Cloud, without process, cannot deliver on the promise of services innovation.

All of the ideas around quickly assembling Cloud-centered applications to support business services simply won't happen without process technology. As we explain later in this book, "Process on Demand" means having the capability to call up Services when needed to change or augment a process that is *already being executed.* This capability is an intrinsic part of the *Service-Oriented Enterprise.*

By taking service process management into the Cloud, services from multiple knowledge sources can be delivered with maximum flexibility and adaptability to meet the requirement that "most services must be customized." On-demand service processes aren't sequenced as in many traditional workflow systems. In contrast they are asynchronous and peer-to-peer, with the high-level process providing the choreography.

Let's turn our attention to the very core issues of providing services. When it comes to service forms of business processes, they reside in the domain of human-to-human interactions. Remember what was cited earlier, "People doing things for other people."

That is to say, service processes cannot be predefined or "flowcharted" in advance. In short, such collaborative human processes are "organic." They represent "emergent processes" that change not only their state, but also their structures as they are born, and then grow and evolve.

Such processes deal with case management and each service renders a unique process instance centered on human-to-human interactions. Keith Swensen writes in the book, *Mastering the Unpredictable* (mtubook.wordpress.com), "The facilitation of knowledge work or what is increasingly known as 'Adaptive Case Management' represents the next imperative in office automation. The desire to fully support knowledge workers within the workplace is not new. What's new is that recent advances in information technology now make the management of unpredictable circumstances a practical reality."

Case management in no way involves the kind of processes you call in IT to analyze, model and code – and get back to you in 18 months with a solution. As Keith Harrison-Broninski writes in his book, *Human Interactions* (mkpress.com/hi), "Human interaction management systems are peer-to-peer, choreographed processes, the kind needed to provide services that delight."

In summary, it's not your father's BPM when it comes to service processes. Is your enterprise ready for the leap from BPM as we know it to Service Process Management (SPM)?

www.mkpress.com

> "Processes don't do work, people do."
> -- John Seely Brown, Former Chief Scientist, Xerox

There's even more when it comes to services processes, something called Social Computing. Instead of hiring or outsourcing armies of call-center employees to meet peak demand, how about employing your customers as experts?

We've already discussed open innovation, but nowhere is the concept more vital than in delivering services. Let's turn to an example of "open services innovation."

Think *"prosumer" (producer–consumer)* and the co-production of value. The role of producers and consumers begins to blur and merge. In a July, 2009 *BusinessWeek* article, Innovation Editor Reena Jena explained, "It's hard to get tangible results from social media. Giants from Coca-Cola to Walmart have set up Web sites where customers can share their interest in the brand. But many of these sites don't attract enough visitors to form a real community or have been slammed by critics, as was the case at schoolyourway.walmart.com. The retailer killed it in 2006 after just three months.

"Unlike many other companies, however, Intuit seems to have figured out a way to benefit from social media. Its insight: Rather than inviting the whole world, the accounting software maker funnels only diehard users of QuickBooks to a site where they can exchange truly helpful information. For customers, that means quicker answers to problems. For the company, this volunteer army means less need for paid technicians.

tinyurl.com/7tnrus4

"Intuit's QuickBooks Live Community is accessible automatically to anyone who opens QuickBooks 2009 on a PC or Mac. The site is similar to macrumors.com or macfixit.com – independent forums where Apple fans can trade tips – except that it's owned and monitored by the company. Intuit chose this 'narrowcast' approach after Chief Executive Brad Smith heard what was going on at the Web site of Intuit's popular TurboTax product. Customers were not only asking technical questions, they were often outshining Intuit's own tech support staff by answering 40% of the queries themselves.

"Since the October 2009 edition of QuickBooks went on sale, traffic on its channel has tripled. At any time, 70% of customer service questions are answered by other QuickBooks owners, says Scott K. Wilder, who oversees the social network. Michelle L. Long of Lee's Summit, Missouri, is often on the site. The 45-year-old accountant has posted more than 5,600 answers.

"The social aspect of the program seems to have helped sales. The Mountain View (Calif.) company has sold 1 million units of QuickBooks at $200 apiece, boosting the software's market share by 4 points, to 94%. All that free tech support is saving Intuit money as well. Wilder points out that since Intuit's community outreach began, 'the number of calls to our customer service lines has been reduced. We don't give out numbers, but there have been cost savings.'"[1]

In response to Jena's article, one reader, Paul, added a valuable insight, "Ingenious in the sense that the only people that can contribute are those that really have had actual exposure to and used the product first hand. By inviting the public to a truly open forum where the advice may or may not apply, a company can hurt its reputation or that of its product if a customer continuously subjects himself to incorrect or inconsistent tips and advice."

Mob-rule constructionism must be managed, especially considering that sockpuppets of your competitors will be ready to pounce. (As we explained earlier, a sockpuppet is a false online identity used for purposes of deception within an online community. Through the false identity, a community member speaks while pretending to be a different person, like a ventriloquist manipulating a hand puppet).

The Social Web can be more than a place where two-way dialogs happen. Moreover, it can also be the place where work gets done. Your *cloud sourced* customers become your service representatives, your call-center in the Cloud, with world-class experts at little or no cost.

Getting There

Services are not just a part of getting a haircut, having your hotel bed made up or a check-up with a doctor. Services also apply to good old physical industrial-age companies, as reported on Dave Gray's blog:[2] "Consider Cemex, a global cement company. What could be more industrial-age than cement? Cement is clearly a product, not a service. And perhaps the most obvious way for a cement company to compete is on price. But to customers, cement is only one aspect of a larger project. Customers don't just care about cement, they want the right cement, in the right amount, at the right place and the right time.

"Cemex wins customers with services like 24/7 delivery, ATM-like ordering systems, education and training for customers, and construction financing. Customers can order online and get text messages when cement is ready for delivery. Cemex will actively manage a customer's cement inventory, to anticipate and respond to demand in real time. Cemex will provide pre-fabricated components like walls, ceilings and basements. And if a customer so desires, Cemex will carefully match the color and texture of older concrete roads and paths."

The Cemex example leads us to change our view from "products" to "products as services." Just consider that *people don't buy drills, they buy holes*. But now the drill is offering us help in drilling that hole. Smart drill icons help you pick the perfect drill setting for each task, so you don't end up drilling too fast, too slow, or with the wrong amount of torque.

Using the term some pundits have adopted, products are becoming *Service Avatars*, delivering services that used to be rendered by humans, if at all.

Then there's that thing we used to call a telephone. You bought one and it let you talk to others. But, now the same "product" is smart. Smart phones can still be used to *talk*, but their real offering of a smart phone is all manner of services they provide.

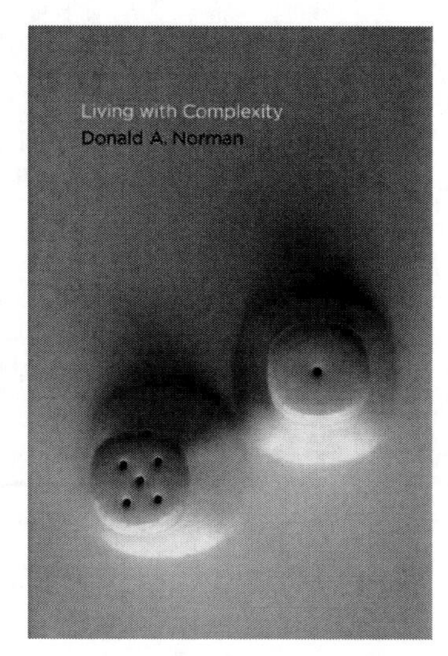

Beyond the obvious smart phone, just about all companies are becoming software companies, embedding digital services into their products. Nike+ is a chip with heels. The shoe sensor's accelerometer measures the amount of time a runner's foot is on the ground, which is inversely proportional to speed. Transmitting at 2.4 GHz, the sensor sends data to a receiver that's attached to an iPod. After the workout, the iPod is synced to a computer running iTunes. Users can access their run history at Nike-Plus.com, browse through a graph that shows all their activity, and then drill down to details about each workout. If they need more motivation, they can enter challenges or set individualized goals, like running 100 miles in a month. Is Nike a shoe company? Or is it a provider of services? It provides a great customer experience. By enmeshing the company into the lives of its customers, the company has now expanded its smart offerings into other product services like the smart, GPS enabled, wrist watch/monitor.

Meanwhile, Ford is a maker of "sophisticated computers-on-wheels." Wi-Fi receivers turn your car into a mobile hot spot; built-in software helps maximize fuel efficiency; ultrasonic sensors enable automatic parallel parking. When the cars are all networked, the possibilities explode. We're rapidly marching toward 2 billion cars, trucks and buses on this planet—time for true innovation. And one key to such innovation is the notion of connecting not just people, but also connecting devices, including automobiles. In other words, services aren't to be rendered to people, but *devices will render services to devices*—Service Avatar to Service Avatar. From your refrigerator to your clothes dryer and appliances of all kinds. smart devices will become a part of the Smart Grid, a key component of a Smarter Planet.

Back to the Smart Car, the most significant feature of next-generation vehicles is *autonomous operation*. The General Motors EN-V can detect and avoid obstacles—including other vehicles—

park themselves and come to you when called by phone. And let's not overlook Google's driverless car. Watch Sebastian Thrun, the director of the Stanford Artificial Intelligence Lab, talk about Google's driverless car.

tinyurl.com/4dnlb9b

Currently, a car spends 96% of its time idle. Cars will spend less time idle. Why would a household buy 2 (or even 3) cars, when they only need 1? Ride to work, then send the car home to your spouse. The operating percent of a car will go from 4% to 96%. Parked cars will be a relic from the past. What happens to car insurance prices if a driver is no longer part of the equation? Perhaps we'll see the rise of *carsitters* or *car nannies* who go along for the ride, for a small fee, to be sure laws aren't broken. And if cars are receiving 20 times more actual use, that would imply that there would be 20 times less cars sold. This is the kind of disruptive change that can reshape the automotive industry.

Applying science to services, as was done in agriculture and goods manufacturing, creates a kind of change that's overwhelming to contemplate. But help is at hand:

- In 1983, bank executive G. Lynn Shostack proposed a design tool called the service blueprint as a tool for service design. The service blueprint connects customer activities and touch points with a company's "front stage" where services are provided, as well as "backstage" operations that support and enable the front stage. *en.wikipedia.org/wiki/Service_blueprint*
- *The Service Research and Innovation Institute,* a non-profit organization initiated by IBM, was formed to lead and support organizations through the massive transformations that will be required. *www.thesrii.org*

Watch IBM's Robert Morris' Keynote at SRII 2011
tinyurl.com/7ljgbfj

- *The Consortium for Service Innovation* is a non-profit alliance of organizations focused primarily on facing the challenges of customer support services. *www.serviceinnovation.org*
- On the practitioner side, the *Service Design Network* was formed in 2004 as an international network of service design professionals, with the purpose of strengthening and developing service innovation practices. Their peer-reviewed publication, *Touchpoint*, is an excellent resource for service design thinking.
  www.service-design-network.org
- The *Information Technology Industry Council (ITI)* is an advocate and thought leader for the information and communications technology (ICT) industry. ITI is dedicated to advocating for its member companies through three main divisions, Environment and Sustainability, Global Policy, and Government Relations. *www.itic.org*
- *The Social Business Council* is a peer-to-peer membership organization of business professionals that are directly involved in planning, leading and executing social business transformation initiatives. *council.dachisgroup.com*
- *The Institute for Business Value* provides thought leadership, strategic insights and recommendations that address critical business challenges and help our clients capitalize on new opportunities. "By understanding weak signals and early indicators of industry transformation, we help our clients to be ahead of the pack." — Peter Korsten, Vice-President and Global Leader, IBM Institute for Business Value. *tinyurl.com/6ymu34c*
- Harvard Business School's *U.S. Competitiveness Project* is a research-led effort to understand and improve the competitiveness of the United States – that is, the ability of firms operating in the U.S. to compete successfully in the global economy. *hbs.edu/competitiveness*

Takeaway

Referring again to GE, Jeffrey Immelt, CEO of General Electric, said, "This economic crisis doesn't represent a cycle. It represents a reset. It's an emotional, raw, social, economic *reset.* People who understand that will prosper. Those who don't will be left behind."

The biggest impediment to service innovation is not a lack of ideas. It's the inability of companies to deliver *services that delight* due to the way they are currently structured. Industrial one-way mass-production logic must be replaced with more reciprocal service logic before progress can be made. This is exceedingly difficult and many companies will not successfully make the transition. It involves changing not only organization structure but the company's dominant culture and logic (as we discussed elsewhere). Peter Senge provides the bottom line for services innovation. In his seminal book, *The Fifth Discipline: The Art and Practice of the Learning Organization,* he writes,

> "The ability to learn faster than your competitors
> may be the only sustainable competitive advantage."

References

---

[1] www.businessweek.com/magazine/content/09_28/b4139066365300.htm

[2] www.dachisgroup.com/2011/11/everything-is-a-service

16. Who You Gonna Call? The Future of IT Services

Technology is revolutionizing the way we live. As new technologies emerge and are adopted by businesses, organizations' information technology services must adapt to help leverage these technologies and take advantage of their potential. No doubt, Information Technology (IT), in our current age of Business/IT fusion *(the business is the system, the system is the business)* and cloud computing is central to business innovation. However, overall internal IT tends to score relatively poorly on the innovation front. Here are some factors that constrain innovation in IT:

- *Strategy:* IT is often weak in having a clearly articulated message on how it supports the company's strategy and differentiation. This can make it hard to focus and evaluate innovative opportunities, which usually results in most ideas failing to gain traction.
- *Process*: Internal IT groups are no different to the majority of businesses in failing to recognize that there is an underlying process to innovation. The absence of a formal disciplined process makes it a real challenge for ideas to successfully make it to market.
- *Measures and Incentives:* The majority of IT shops have incredibly basic performance measurement—with the primary focus on annual cost savings. There is virtually no mechanism for tracking investments and value creation over periods of longer than one year. Furthermore even annual "investment" budgets typically get assigned to the largest and most mature application areas rather than towards emerging opportunity areas.
- *Market Intelligence:* Too few IT shops have meaningful engagement with their company's market and competitive intelligence. This means they are not being fed the business insights about new opportunities and risks. This results in IT being too comfortable with the status quo.
- *Fast Followers:* A lot of the innovation that does exist in IT revolves around expanding technology areas to cover faster, better, cheaper ways to deliver what it is already doing. It rarely encompasses strategic innovation, business process innovation, business model innovation. As a consequence, IT innovation rarely creates significant differentiation and in effect IT becomes "followers" of the latest industry analyst report on best practice.
- *The Bill Back and Cost Recovery Model:* Internal IT shops are still fundamentally built on the individual project cost recovery model. This acts as a (subtle) disincentive to improving productivity and creating opportunities for leverage and synergy. Also as described in the Measures and Incentives point above, it is difficult to introduce different business and process models.

The 800-pound gorilla in the room is the inescapable fact that the economics of scale alone dictate that services derived from the Cloud will typically be less expensive than any that an internal IT shop can provide.

Does your company generate its own electricity? Likewise with basic IT services. Now compound the issue with how much an internal IT shop can allocate for security, compliance, disaster recovery, 7x24x365 operations, and it is quickly apparent that the economics dictate, with very few and rare exceptions that internal IT will evolve the way of internal electricity, internal shipping, internal telecommunications, and pretty much any other aspect of the business that doesn't directly create value for customers or competitive differentiation for the company. All of these resources will be sourced from a more proficient provider. Many of the objections to that concept are effectively, in the long term, red herrings.

As they did when complexity grew in the enterprise use of IT way back in the late 1950s, companies will reach outside their internal IT departments and call in outside IT services companies that have assembled top-tier technology talent in cloud computing. Over the next few years, we will see some key trends developing that will change the IT services industry as a result of the emergence of cloud computing.

What is the IT services industry?

In the beginning, way back when this newfangled thing called a computer entered the enterprise, life was kind of simple and "data processing" basically involved computer programming. But as the uses of computers grew in both breadth and depth within the enterprise, things quickly became complicated. 1959 marks the year when companies looked outside to tame the growing complexity of enterprise computing. That was the year that what is now today a major "systems integrator," Computer Sciences Corporation (CSC), was founded. CSC was founded in El Segundo, California in April 1959 by Roy Nutt, Fletcher Jones and Bob Patrick. A systems integrator is a company that specializes in bringing together component subsystems into a whole and ensuring that those subsystems function together. Today, CSC employs about 93,000 people in 90 countries and ranks among the largest outsourcing companies in the world and has been a Fortune 500 company since 1995.

But CSC isn't the only IT services player in the game. IBM Global Services (426,751 employees in 2010), Dell Services (formerly Perot Systems), HP Enterprise Services (formerly EDS), Logica, Accenture, BearingPoint, Capgemini, Deloitte Consulting, HCL Technologies, Infosys Technologies, Siemens IT Solutions and Services, Tata Consultancy Services, and Wipro. You might consider these firms as a source of "timesharing" of technological and engineering talents that individual companies could ill afford.

To see how the IT services providers will develop in the Cloud era, we must first work backwards from the end state—the "consumerization" of IT. Vizio has announced it will be delivering an Android phone and tablet; LG and Samsung already support HTML5, and have apps available on their televisions; and you can't sit in a plane and not see iPads everywhere (sometimes even running virtual desktops over the plane's WiFi).

A key for any company's IT services providers is how to support, manage, and deliver value to these diverse devices. The other part of this issue is that we now have multiple devices per person. Thus IT services have to move from supporting, managing, and delivering value to *a device*, and start delivering it to an *entity*, while maintaining context across time, location, and devices.

As this environment develops, we are starting to see what some are calling the "pervasive Cloud," or as Kevin Kelly calls it, "the One." One part of the pervasive Cloud is that these very capable, generally mobile devices "integrate" across multiple cloud offerings, giving the illusion of a single, pervasive Cloud. Many smart phone apps do this today, for example. This integration will take place at two levels: one at the actual device, the other behind the scenes via cloud brokers and integration tools and services —the Intercloud.

All this connects to another change in capabilities that is increasingly in demand: unified communications and messaging. Connecting people and their personal devices and formats means accommodating the available form factors for each and every participant, as the imposition of a single standard across not just one enterprise but across the entire globe is clearly impossible.

The ubiquitous high-bandwidth Internet connections running Internet Protocols, the increasing range of devices to make use of this capability, and the shift in focus from machine-to-machine networking toward people and collaboration have resulted in a complex environment. Many, if not most people today, use more than one device. These can be a combination of a work PC, a separate home PC, a smart phone and a tablet PC.

Most large enterprises grow organically, usually in an un-planned manner with no serious attempt at optimizing the use of corporate bandwidth and connections. In fact, the budgets to operate all of the devices and services may well be held in different operating units with no overall view of the total costs.

Unified Communications or UC is the integration of all of the communication connections and devices into a single cohesive environment. It presents a consistent service to all devices from a cost-effective centralized resource. Even though it's technically inaccurate, comparisons have been made to this being a kind of virtualization for communication devices. However, in the context of optimizing resources, there are similarities.

Unified Messaging or UM is concerned with creating a *common interface, or experience for users* to hide the differences between various services running on different devices at different locations. One of the fundamental goals of UM is to ensure that any messaging service, such as voice, email, or instant messaging can be used from any device, regardless of format or location, fixed or mobile.

In practice, it is rarely possible to separate UC and UM as the two aspects are largely interdependent and most products combine them. The term "Presence" is used to indicate the ability of a UC and messaging service to understand the current location of the user in terms of the device being used and the services that can be supported.

At the Mobile World Congress 2010 in Barcelona, Google's then CEO Eric Schmidt explained that Google is moving forward with a "mobile first" mantra. "It's like magic. All of a sudden you can do things that it never occurred to you were possible. Our programmers are working on products from a 'Mobile First' perspective. That's a major change. Every recent product announcement we have made – and of course we have a desktop version – is being made from the point of view of it being used on a high-performance mobile phone on all the browsers that are available. Our programmers want to work on apps for mobile that you can't get on a desktop – apps that are personal and location aware."

Another IT services opportunity is in the creation and integration of Cloud services, either on the mobile device or on the back-end systems. It is likely that Platform-as-a-Service (PaaS) will be the foundation going forward for new traditional application development activities, for the integration platform behind the pervasive Cloud, and for the creation and delivery of application components. This is because PaaS abstracts hardware and system software concerns away from the developers, so that they can focus on business information, process, and structure rather than technical implementation details. This allows both faster development and a broader spectrum of potential developers—think logical, disciplined MBAs and "Shadow IT."

One interesting consequence of the pervasive Cloud is the increasing connectivity of technology as more and more devices are connected to the Internet and make their information and capabilities available online (the Internet of Things). At the NASA CIO conference in August 2010, Vint Cerf presented the things Google was working on, including a true multi-path Internet protocol that does not require ubiquitous connectivity to work, which NASA is adopting as part of its Deep Space Network. Another of his examples was automated wine cellars, which might be a California thing, but which nonetheless demonstrate the possibilities. Many medical devices are moving in this direction, such as stay-at-home medicine with pill packs that know if patients have taken their meds and responds to radio frequency identification (RFID) requests to report.

All forms of direct data input from devices, including RFID, and their support will be a major IT services opportunity for the future, not only with data collection and processing, but in the "system management" aspects as well, include provisioning, administering, securing, and monitoring resources in the Cloud. The key bottlenecks have been the cost of the chips, which are coming down rapidly (some companies are even experimenting with actually printing simple forms of the chips), and the cost of communications like transponders and antennas for discrete or item-level RFID inquiry response, which is also being addressed quickly.

This increasing connectedness in turn creates the "Big Data" problem, where there is too much information to be effectively understood by humans in a traditional reporting manner, which then leads to Business Intelligence opportunities for IT services. There is also a Storage-as-a-Service (SaaS) play here as well, if the economics and performance issues can be addressed. As storage becomes increasingly hosted, density rather than functionality will drive purchase decisions. The type of BI this Big Data problem requires is different than traditional BI. First, it must be embedded into the actual data stream, maybe as part of a composite application. Then it will be expected to be practically employable (responding to what is happening) and therefore somewhat more predictive-oriented than reporting-oriented. BI systems such as Splunk (www.splunk.com) will become typical. Why? The IT systems and infrastructure that run your business generate massive volumes of data every millisecond of every day. This data contains a definitive record of all user transactions, customer behavior, machine behavior, security threats, fraudulent activity and more. It's also dynamic, unstructured and non-standard and makes up the majority of the data in your organization. Making use of this data requires a solution that understands this data—one that can collect, index and harness massively diverse and dynamic data types, without limits for operational visibility, proactive monitoring and business insights.

Between here and there is a transitional opportunity: Cloud *appliances*. It is likely that IT departments will be slow to give up control. Appliances provide the opportunity to configure IT for 100 percent utilization, and then if the appliance is linked back to a Cloud for burst or end-of-period exceptional consumption, the fixed cost/variable cost balance is restored.

Underlying all of this are the cloud computing technologies, integrative client technologies (e.g., HTML5), and the ability to integrate and broker multiple clouds. This is a real opportunity for a new breed of IT services companies, Cloud Brokers, in the areas of information governance, security, compliance, and risk management.

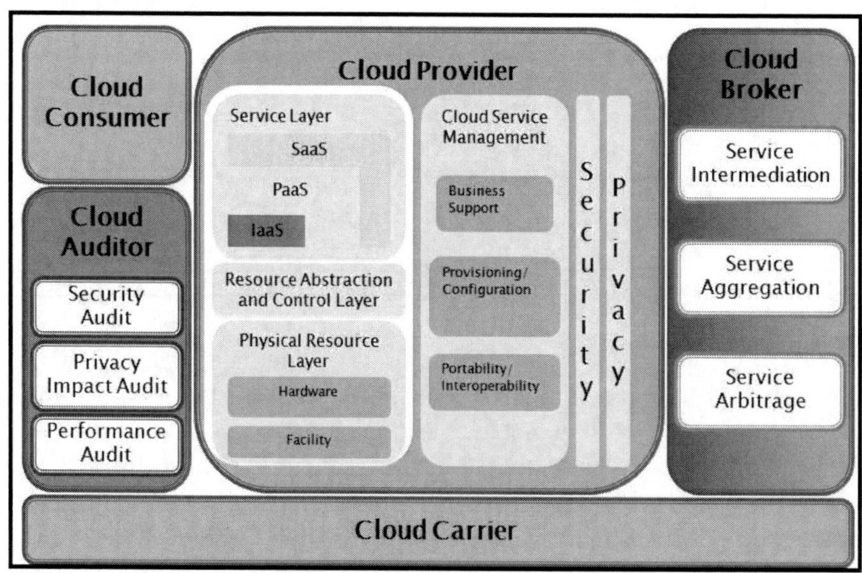

NIST Reference Architecture

New laws and regulations at the national and state level and the extension of existing rules are creating a significant pain point for companies. In the same way that companies gave up doing their own payrolls because they couldn't afford the number of accountants and lawyers necessary to keep up with all the rules, data in general is about to enter into the same environment. This creates a huge opportunity for advising, implementing, and taking over this function by Cloud Brokers.

Current security technologies and offerings are not well engineered to support the growing needs of operating in the Cloud. Traditional boundary models (once you are in, you are in) and binary models (you either have access or you don't) won't work long-term. The Carnegie Mellon University's Computer Emergency Response Team (CERT) and CERT Coordination Center (CERT/CC) has said that the rate of new malware evolution will exceed the ability of security vendors to keep up. This means three technology opportunities: digital rights management (DRM), policy management to implement DRM, and self-protecting data models such as those being discussed in the Jericho Forum of the Open Group.

Looking beyond the pure technology perspective, IT has been undergoing a continuous evolution to greater and greater levels of abstraction. Today virtualization's goal is to totally abstract the hardware and system software. The logical end state would be the abstraction of IT itself. We are starting to see this happen: The Corporate Executive Board's recent report on the future of IT essentially predicts that most of IT will be either merged into departments, be integrated into a central, shared services function, or be effectively outsourced—not to a technology provider, but to a provider of business services that will bring along its own technology for the operational aspects of the information. This would suggest that the long-term opportunity for IT services is to supply services to the emerging Business Services Industry—not just Business Process Outsourcing (BPO) but also Business Services as a Service (BSaaS).

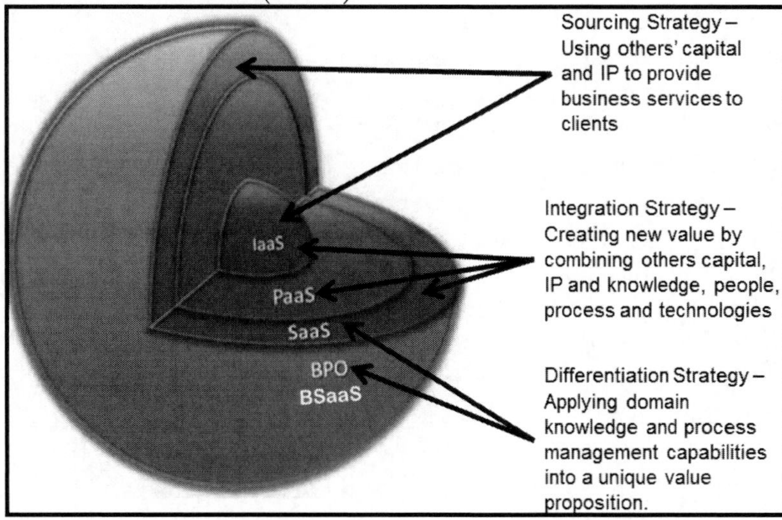

Clearly, mobility will be the catalyst for the technologies that emerge in the Cloud that can enable anywhere, anytime, any-device computing (and maintain context across every device); and that can manage and support anywhere, anytime, any-device computing.

Under this assumption, information security, governance and compliance, and risk management technology will dictate the boundaries and stage gates within which, at least at the enterprise level, the desired capabilities will operate. One could almost imagine a new Microsoft providing the DRM, policy management, and self-protecting data environment, though one would hope that "open standards" will get there first. There will likely be much risk management discussion of capabilities, as consumers and startups eschew the conservatism of traditional larger enterprises and provide capabilities outside of the existing parameters of information protection.

Within these constraints, technology that enables the choreography and management of end-to-end business processes will be a vitally important technology provided by both Cloud service providers (CSPs) and under the covers of the business service providers (BSPs).

Social analytics will also emerge as a dominant technology service sought by customers. This one is easy, as it is a natural for one of the aforementioned business services. Look at the pattern of usage for Google Analytics. This could be considered a special case of the Big Data discussion above (and the resulting need for embedded, predictive analytics), but the specific domain knowledge and specialized skills needed to utilize Google Analytics has driven most of this work to niche business services providers (in the same way software-as-a-service will be narrow-focused and embody "deep-knowledge"), creating a long-tail market and business model. Again, the IT services opportunity will be to serve those business service providers.

The geographic spread of technology services cannot be ignored. During the past decade, the U.S. and Europe were the largest markets for IT services. Many think this decade will see the emergence of new markets such as Africa, India, and China.

This is an interesting debate, but it is the wrong way to think about the future. Most businesses have organized around geographies because of the physical and economic frictions involved in centralized services and the difficulty of tailoring to geographically localized requirements. The Cloud eliminates many of these frictions as part of a process that the Internet started.

While the ethnic and cultural preferences and prejudices that dictated a lot of regionalization models still exist, they are not geographically unique anymore. The world, not just the U.S., is becoming a melting pot, so again calling on Anderson's long tail, the larger markets are likely aggregations of geographically dispersed smaller niche markets that have much more in common with their global aggregation than their regional proximity.

Tie this with the radically increasing urbanization, and the right question is not what region or country to invest in, but what city to invest in. This is clearly seen in China, where Tier 1 and 2 cities have the growth and infrastructure to support IT. There are cities in Africa, such as Cairo, Lagos, Kinshasa, Johannesburg, Cape Town, Abidjan, and Casablanca that may be better opportunities for investment in IT services than some places in the U.S. or China. Likewise, we are seeing growth in Europe along the lines of cities rather than countries, particularly in Eastern Europe. There are also surprises, such as Chittagong, Bangladesh. It is likely that in 10 to 20 years, economic zones will really be organized around cities rather than countries or regions. In fact, we are already seeing some of this as China emerges. Some of these key future markets include:

- China: Chengdu, Suzhou, Nanjing, Chonqing
- India: Ahmedabad, Chennai, Bangalore
- South America: Santiago, Campinas
- Europe: London, Spain (Mallorca, Pamplona, Badajoz)
- Middle East/Africa: Tel Aviv, Abu Dhabi, Cape Town
- North America: Raleigh-Durham, Austin, Salt Lake City, Calgary
- Asia/Pacific: Melbourne, Kuala Lumpur, Hanoi

The IT services industry is poised for change at a fundamental level, reflecting the ongoing revolution in how we use and deliver technology. These trends, and ones we cannot even fathom yet, will present numerous opportunities for IT services. We have to be very careful not to fall into the trap of thinking in traditional ways about these opportunities. In his December 2007 speech at the Entertainment Group conference, Kevin Kelly (publisher of *Whole Earth Review* and executive editor of *Wired*) talked about how in the beginning everyone thought the Web would be like TV, only better.

Well, it wasn't. Things like Wikipedia, Google, and Facebook weren't even imagined. The issue was traditional economic model assumptions that no longer held with consumers versus corporate-driven IT, different economic frictions, and enablers. As Kelly said, those types of businesses were deemed impossible in theory, but turned out to be possible in practice. The IT services industry must make sure that it takes advantage of these opportunities to deliver new services and value to its customers—and that will require really examining the assumptions of what is possible.

The Rise of the Cloud Broker

From all of this will come one form of innovation that will enable many more – the rise of the Cloud Brokers (briefly introduced in Chapter 10). And not just brokers of IT services, but entire business service offerings available in the Cloud.

According to Gartner, through 2015, Cloud Services Brokerage (CSB) will represent the single largest revenue growth opportunity in cloud computing. But like so many areas of cloud computing, there is confusion about both the terms and the very nature of this concept.

First, there is not just one type of Cloud Services Broker. As ubiquitous or utility computing evolves, there will be multiple types of brokerage services appearing, in the same way there are multiple types of clouds. For example, Amazon, Google, and Rackspace all use different underlying (and basically incompatible) technologies and approaches. There are also multiple "as a service" types of offerings—such as Infrastructure as a Service, Platform as a Service, and Software as a Service—and all of those providers are diverse and incompatible. And that's just the tip of the iceberg. Cloud computing as a technology platform is varied, evolving, and diverging as services become more nuanced and specialized.

The brokerage market will be just as diverse as the cloud is. There will be brokers that do nothing but perform a management and integration service across multiple clouds and Cloud vendors for infrastructure and platform needs. The cloud computing landscape is evolving rapidly, with more and more players introducing Cloud products and services of all kinds.

Most recently we've seen the announcements by VMware partners, including Terremark, BlueLock, and others, as well as the introduction of Rackspace's Cloud Servers. EMC is planning to offer a compute cloud, in addition to its existing Atmos storage cloud. And Dell is taking a broad approach through partnerships with Microsoft (Windows Azure Platform appliance), VMware, and by helping to develop Open Stack.

As the proliferation of offerings continues to accelerate, IT managers have questions about how to proceed. How can you evaluate the range of potential Cloud offerings to find the right match? How do you route an application or workload to a target cloud and make sure that it works? How do you integrate it with IT applications running back in the datacenter?

Even within a single cloud, deploying an application requires learning the provider's operating environment, management tools, and business terms and conditions. Doing this for every cloud provider you may wish to utilize is likely to prove daunting and not cost-effective. In a cloud computing environment characterized by multiple providers, each with its own service terms, operating platforms, management systems, security levels, and disaster recovery approaches, the specialized expertise and value-add of a Cloud Services Broker will help IT managers find the right Cloud offering, deploy their application in the Cloud, and manage it properly.

Cloud Brokers will also form around domain areas (pharmaceuticals, IT, soft goods retail, process manufacturing, and so on) or applications areas (supply chain management, enterprise resource planning, human resources, payroll, and so on)—perhaps even combinations of these—so that they become shopping centers from which companies can assemble the necessary services for their businesses.

In a similar vein, we can expect the rise of cloud services brokers that specialize in Big Data—a consequence of the evolving connectivity of things and the concurrent big data deluge. These brokers will specialize in the collection and storage of domain, industry, or applications area data and the associated analytics. An example of this could be a clearinghouse for point of sale (POS) data collected across retailers and made available to manufacturers.

It is likely that the evolution of Cloud Services Brokers will be the logical extension of current business process outsourcing services (BPO) to include more end-to-end and higher-level knowledge skills incorporated with applications (or application components rendered as Cloud services) and data. One can easily imagine everything that's not core to a business, or not directly creating value for a business's customers, eventually being outsourced. For example, the simple process of hiring someone could be done completely virtually by using multiple applications, databases, and external services via the Cloud.

We even suspect there will be a transitional market opportunity for companies to sell excess capacity on their internal systems via cloud brokers, especially as they move their hygienic and housekeeping systems to the Cloud. You already see this with Toronto-based Enomaly, which lets companies buy and sell unused cloud computing capacity in a clearinghouse called SpotCloud. The service also gives companies the ability to switch services on demand to get the best price while still receiving a single bill.

And lastly, there is the traditional broker—companies like Sterling Commerce—that can route purchases or service requests to the "best" provider from among the variety of offerings. Think of it as "least-cost routing" for cloud computing. The benefits of Cloud Services Brokers using specialized tools to identify the most appropriate resources, and then map the requirements of an enterprise application or process to those resources, will soon become apparent. Cloud services brokers will be able to automatically route data, applications, and infrastructure needs based on key criteria such as price, location (including many legislative and regulatory jurisdictional data storage location requirements), latency needs, SLA level, supported operating systems, scalability, backup/disaster recovery capabilities, and regulatory requirements. IT and business process managers will be able to run applications or route workflows where they truly belong, while the broker takes care of the underlying details. For example, if you are hiring an employee in India, some of the BPO activities might be routed to a different service provider than if you were hiring someone in Russia or China or the United States.

But wait. Given that the notion of a Cloud Services Broker is about as well defined as a marshmallow, what can we say about them?

Gartner is most likely right to forecast that the market, however it evolves, will be huge. The Cloud has taken so much friction out of the economics of IT that a long tail will be in full effect, with countless different service offerings to be evaluated, chosen from, integrated with, managed, and so on—and companies will gladly pass that pain off to a broker.

Meanwhile, the Object Management Group with its Cloud Standards Customer Council is currently hard at work to be sure that the "voice of the customer" is heard by the standards bodies in the IT industry as they take on the big issues of interoperability.

WHAT'S NEW

Peter Fingar on Cloud Computing

In this podcast Peter takes us on a fascinating journey to explore what the Cloud portends for business. In the past, IT was about productivity; now in the Cloud it's about collaboration, a shared information base, and collective intelligence.

Learn More ▸

cloud-council.org

Whither the Internal IT shop?

As Boyd learned, the structure of the organization dictates its ability to practice OODA. The key is to eliminate distractions. Do only those things that create value for your customers, or for your organization or creates separation between you and your competitors; then do nothing that can be better done by someone else.

In general this means that the structure and organization of IT is changing under our feet.

In short, about 25% of what would traditionally be in IT is moving to individual business units and departments. It is the late 70's and early 80's all over again, when we did not know a department had a PC until we got the call asking for a report from the general ledger system. Well now it is not PCs, it is Software as a Service applications.

"The Long Tail" was popularized by Chris Anderson in an October 2004 *Wired* magazine article. The basic idea is to realize significant profit out of selling small volumes of hard-to-find items to many customers instead of only selling large volumes of a reduced number of popular items. The total sales of this large number of "non-hit items" is called the Long Tail.

Because the Cloud has taken so much economic friction out of the software marketplace, we are seeing the rise of a Long Tail where increasingly specific high functionality, narrow breadth, deep domain knowledge applications are rising up to serve unmet needs with extremely high payback, but could never have been created as traditional software or as one off applications inside of IT due to economics. Think of this as the individual departments maximizing their own OODA loops, which a central IT function could not do for them.

Companies are also reverting to shared services models of organization. If an organization or a department does not directly create value for customers, or provide some form of competitive differentiation, separation or advantage, then it will likely be rolled into a shared services organization. Once traditional silos of capability are within one organization, then traditional stovepipe business processes and associated applications begin to disappear, replaced with horizontal business transactional systems. Why do I need to enter an HR transaction, a facilities transaction, an IT transaction, a Payroll transaction, a benefits transaction, etc? I should simply enter a new employee transaction and the shared services organization makes it work. Once this begins then opportunities for new systems, Hollywood model services and even business process outsourcing begin to appear, further reducing the internal IT workload related to non-core processes.

Last, what value is created for your customers by you running a network or storing data, or managing UPS? Honestly, little to none? Few companies are in the facilities business anymore, nor generating electricity, nor running their own internal phone companies. So more and more IT infrastructure will be outsourced. Why? Reasons include scale, the rarity of trained staff, and the elimination of distractions, enabling the many other ways internal IT staff members can truly focus on business outcomes and contribute their general systems thinking skills to the business.

With reductions in managing infrastructure workloads, internal IT can focus on creating customer value, in creating competitive differentiation, becoming a single point of security, compliance, provisioning, management, monitoring – integrating the many SaaS, business service providers and external infrastructure providers into a seamless whole for the organization – and keeping, operating and maintaining those things that cannot be let outside the enterprise.

The internal IT organization becomes the focal point for working with external Cloud Brokers—an internal broker of Cloud Brokers if you will. Thus the internal IT organization becomes a procurer, an orchestrator and a choreographer for the IT services derived from the Cloud. This provides the framework and foundation needed to enable true business innovation.

In August of 2010 the Corporate Executive Board released its "The Future of Corporate IT" study (updated October of 2011). A summary of it can be found at the link below. The report opens with a backdrop, then goes on to describe Five Radical Shifts in IT Value, Ownership, and Role. A brief summary each is provided here.

tinyurl.com/765ajsd

"In 2005, less than 25% of business leaders rated their organization's IT function effective at delivering the capabilities they needed. Today the number hasn't changed. IT functions have strived tirelessly to understand demand, set priorities, deliver effectively, and capture value, yet the results still disappoint. Business and IT leaders alike feel they should be getting more—more efficiency, more innovation, more value—from technology.

"But there are still unasked questions. Among all the talk of engagement, alignment, and 'being part of the business,' one assumption is never challenged—that for information technology to grow in strategic importance, so must the IT function. But what if this is not the case? What if a dedicated, standalone IT function is no longer the best option and the function's resources and responsibilities were better located elsewhere? Our work revealed Five Radical Shifts in IT Value, Ownership, and Role that make these questions necessary and urgent:

1. *Information Over Process:* Competitive advantage from information technology shifts toward customer experience, data analytics, and knowledge worker enablement. Thus information management skills will rise in importance relative to business process design.
2. *IT Embedded in Business Services:* Centrally provided applications and infrastructure will be embedded in business services and delivered by a business shared services organization.
3. *Externalized Service Delivery:* Externalization of applications development, infrastructure operations, and back-office processes continues, eroding the "factory" side of the IT function. The pace will accelerate as the Cloud enables the externalization of up to 80% of application lifetime spending. As this occurs, internal roles will shift from being technology providers to technology brokers.
4. *Greater Business Partner Responsibility:* Business unit leaders and end users will play a greater role in obtaining and managing technology for themselves where differentiation has more value than standardization.
5. *Diminished Standalone IT Role:* IT roles will embed in business services and evolve into business roles, or be externalized. Remaining IT roles will be housed in a shared service business group. The CIO position will expand to lead this group or shrink to manage IT brokerage, procurement and integration.

Importantly, many staff members in IT organizations are far more trained in general systems thinking than specialists in marketing, finance and other company departments—and systems thinking is the "core" core competency in today's complex business world. Thus, IT staff who become business savvy will likely assume new leadership roles in the organization.

Takeaway

Not only does the Cloud create opportunities for a new breed of organizations in the IT services industry, it portends a reinvention of the role of internal IT services organizations. To repeat an observation we made earlier, the internal IT organization becomes the focal point for working with external Cloud Brokers—an internal broker of Cloud Brokers if you will. Thus the internal IT organization becomes a procurer, an orchestrator and a choreographer for the IT services derived from the Cloud. This provides the framework and foundation needed to enable true business innovation, and sets the stage for new generation of CIOs, as discussed in the next chapter.

17. The Future of the CIO

If mankind survives, by 2050 the world's human population will see the equivalent of three Chinas being added. The changes in the bigger world outside the business world are so immense that what it means to be a business is changing on a scale never before imagined. The physical world is merging with the digital world, and the role of what we now call the CIO will change, and change significantly.

Setting the Stage for a CIO rEvolution.

With the complete fusion of technology into the modern enterprise, technology and business have become inseparable. The technology is the business; the business is the technology. Together as one, they are needed to address the bigger world, the bigger society, in which a business must operate. And it's that bigger world that has changed as a result of the hyper-connectivity of the Internet that, in turn, has given rise to total global competition and Social networks where the future is being discussed, debated and transformed.

It's in this context that the role of the CIO comes into question. From compliance to cloud computing, from budget cuts to Social Networks and business innovation: it is clear that the CIO is living in turbulent times.

What will be on the CIO's agenda ahead? Will there actually be a CIO in the future?

Bear with us for a minute for a little history is in order to see where we stand in the evolution of technology roles today. We've seen many step-changes in technology over the past 50 years, from mainframes to minicomputers to PCs to client/server enterprise networks, and on to the Internet. We've also seen major shifts in how we command computers to do what they do, from board wiring of tabulating machines to machine-language programming to procedural programming to object-oriented programming of various flavors and granularities now including Web services. We've witnessed the shift from monolithic to modular and client/server architectures and on to service-oriented architectures. We've seen changing IT and IT leadership roles, from board wirer to programmer to systems analyst to information architect to EDP manager to IT director and on to the CIO. Today, data centers are staffed by systems analysts, database administrators, database programmers, data administrators and technical systems administrators. Quite an expensive command-and-control army has grown up to support and control central IT where a company's systems-of-record are housed.

Enter stage left, the read-write Internet, Web 2.0, where just about anybody can go beyond consuming information to producing it as well. Programmer? Who needs one? You just access pre-programmed services to create a Web site, join Facebook, tweet on Twitter, update a Wiki, create a blog or mashup your custom Google apps. Just about anyone can do it, just as they could with a spreadsheet—no central IT department is needed. Web 2.0 represents the consumerization of IT, and you might think that's the end of this one-minute history of technology. But, no, all that was yesterday.

Now the real action has just begun with the read-write-*execute* Internet—the Cloud. If you take all of the amazing advances in computing over the past 50 years, as described above, the Cloud represents the knee in an exponential growth curve, making cloud computing the new baseline for business and human collaboration models we have yet to conceive of, and the new Bill Gate's and Jeff Bezo's we have yet to hear of.

We'll repeat it again: in the past, information technology was about productivity; now it's about collaboration, a shared information base and collective intelligence—the wisdom of crowds, social networks and Cloud sourcing of computing power, all in the hands of everyday people.

Remember that mechanically inclined board wirer or machine-language programmer? Or that EDP manager, or that IT director or that CIO? Move over, for it's now time for the chief cloud officer (CCO)— and no, we really don't expect this title to catch on.

The role of the CCO is to provide leadership and guidance in a brave new world where not just programming but also technology infrastructures are abstracted as services—Everything as a Service (EaaS). It is basically a matter of upping the level of abstraction that IT professionals will have to embrace and master. At each step along the way, some IT professionals have always found themselves not suited to the new levels of abstraction, and faded away.

Each step represented a climb from a machine abstraction to a business abstraction, and business process management represents a quantum leap to the *business domain* as the central abstraction of software development. Perhaps the highest level of abstraction is that of "general system thinking." As W. Edwards Deming, the father of the quality movement pointed out, it is "the system" that is the problem. End-to-end business processes are dynamic systems, but today's business professionals are generally not trained in general systems thinking. Too often constrained by a perspective limited to ingrained business practices, rigid scripts, and structured input-output work, few professionals have a wide-angle view of, or experience dealing with, end-to-end business processes.

The worlds of business and technology are growing more complex and managing that complexity is the goal of systems thinking. It focuses on the whole, not the parts, of a complex system. It concentrates on the interfaces and boundaries of components, on their connections and arrangement, and on the potential for holistic systems to achieve results that are greater than the sum of their parts. Mastering systems thinking means overcoming the major obstacles to building the 21st-century enterprise – for every enterprise business process is a whole system.

Although tech-savvy, the CCO is all business, probably coming out of the ranks of operations or an extremely business-savvy CIO. It will indeed be informed leadership, not command-and-control management of computing and information resources that will shape the future of companies and countries in the current era of global economic crisis and unexpected change. Agility is no longer, a *nice-to-have* option. It's the entry price. Lead, follow or get out of the way.

There is much to learn and cultural barriers to overcome, but the company of the future will not be the company of today. The future is here now, as we shift from information technology (IT) with the focus on productivity to business technology (BT) with the focus on collaboration; and as we shift from systems-of-record to systems of boundless collaboration backed by endless computational resources available to all.

Contemplating any company's transformation to adapt to the changed world, the implications for IT professionals are profound. Companies will need a far greater contribution from IT than ever before, but that contribution will be of a substantially different nature. For the adaptive enterprise to come about, a system-wide view of the company is needed, and IT professionals have such a view, far more than the marketing, legal, financial, and other specialists in the firm.

Building the process-managed, real-time, "social Enterprise" will demand innovation centered on the discipline of "general systems thinking" from a new generation of IT professionals, stressing some to their limits as the cloud computing and Social Networking paradigm shifts take hold. It is not your father's IT shop any more. Multidisciplinary skills now outweigh yesterday's technical skills. BPM skills give way to social BPM skills. Data center management skills give way to cloud brokerage management skills.

You Can't Judge a Job by Its Title.

Just as today's popular title of Chief Information Officer (CIO) evolved from that of programmer to EDP manager to IT director, get ready for a new title that reflects the new skill set needed for leadership as companies go beyond IT and embrace business technology (BT). In a cover story in *CIO* magazine in 2005, Sue Bushell, argued for the transformation of the CIO to the CPO or Chief Process Officer, for it is end-to-end process management that companies want, over and above just information management.

tinyurl.com/7skaw3g

Now process management must extend across the entire value chain in the Cloud and provide social BPM and social CRM capabilities. So to indicate the new role and skill set of the CIO, here are some titles that may better reflect the job at hand:

- CCO- Chief Cloud Officer
- CPO- Chief Process Officer
- CDO- Chief Digitization Officer
- CDC- Chief Dot Connector
- CIO- Chief Innovation Officer
- CEO- Chief Executive Officer

Hmm? About those last two titles? What skill set is needed by a CEO? It is the responsibility of the chief executive officer to align the company, internally and externally, with its strategic vision. The core duty of a CEO is to manage the boundaries and facilitate business outside of the company while guiding employees and other executive officers toward a central objective. A CEO must have a balance of internal and external initiatives to build a sustainable company.

Ditto for the next-generation CIO, whatever the title.

It's all about mastering the unpredictable.

No, CEOs and next-generation CIOs won't be the ones who do all the weak-ties work. Instead they will be the ones who provide the environment, guidance and the tools for knowledge workers throughout the workforce to capitalize on weak ties. This also means turning decision making upside down, at the bottom of the organizational pyramid, with front-line staff where actual events take place. Is this avalanche of technology and change in expectations and working practice really to be something made to work in the favor of an enterprise to actually create more sales and value? As a reality check there is the well known example of how Walmart, one of the savviest retailers, has gained by supporting and enabling its front-line staff to respond to in-store events and make quick decisions that boost its sales.

Walmart has driven many new technologies through the hype cycle and into real value ahead of others. The Telxon is a hand-held bar-code scanner with a wireless connection to the store's computer. When pointed at any product, the Telxon reveals astonishing amounts of information: the quantity that should be on the shelf, the availability from the nearest warehouse, the retail price and most amazing of all, the markup. All employees are given access to this information, because in theory at least anyone in the store can order a couple of extra pallets of anything and discount the item heavily as a Volume Producing Item (VPI), competing with other departments to rack up the most profitable sales each month. Floor clerks even have portable equipment to print their own price stickers. This is how Walmart detects demand and responds to it by distributing decision-making power to the grass-roots level. It's as simple, yet as radical, as that.

One employee recounted the story of test-marketing tents that could protect cars for people who didn't have enough garage space. They sold out quickly and several customers came in asking for more. Clearly this was an exceptional case of word-of-mouth marketing, so the employee ordered a truckload of tent garages, "Which I shouldn't have done really without asking someone," he said with a shrug, "because I hadn't been working at the store for long." But the item was a huge success. His VPI was the biggest in store history – not bad for a new employee.

Today's Realities

The book, *Enterprise Cloud Computing*, discusses the role of IT funding models. Funding Models are at the heart of weathering current economic conditions for IT. In 2009, an open-ended survey was conducted that focused on how CIOs felt in the current economic circumstances, and what role IT was playing in their organizations. As with all surveys, the questions inevitably influence the outcome somewhat.

For example, if you ask a CIO if costs are important then you'll struggle to find one that says "no." But the open-ended survey led quite naturally into a conversation about how well the IT role was being played. Should, or could, IT be playing other roles? And perhaps most crucially of all, the survey led to a comparison between how IT was used and how well their organizations were weathering unexpected change in the global economy. How does the CIO see her role? The results showed three common profiles of the CIOs surveyed:

- Technology Utility (24%) = IT is managed as a pure utility
- Service Center (39%) = IT assets are packaged to provide specific services
- Business Technology (37%) = IT is a key asset in the leadership of the business

What lies behind these headlines is really interesting as 490 CIOs effectively ended up comparing notes on what, and how, things are working, or not working. The encouraging part is that a third of the CIOs now think the credit crunch has driven a reappraisal of how technology can genuinely move to be a revenue, margin, or performance enhancing part of the business model.

But can this really happen without attention to the funding model? IT has traditionally been a *back office* tool designed to centralize and improve key business processes and as a result, improve enterprise efficiency while reducing costs. As such IT has been treated as a "business cost" funded through the annual budgeting cycle as an overhead to be recovered. There are various ways to apportion this overhead, but at the end of the day it is a cost to the business, and like all other overhead, needs to be hammered down.

The pillars of an IT investment have been to invest a relatively large sum of money in a long project cycle, and then wait for a payback by gambling on the stability of the situation. At the end of a given IT investment cycle an enterprise should have a permanent competitive advantage. The end result of this approach is that the ongoing costs of IT have increased so much that the headroom in the budget for new investment continues to decrease. When funding is in short supply and stability non-existent, the traditional pillars for starting new IT projects are not generally acceptable. So CIOs should stop fooling themselves. In reality, much of the IT estate is a requirement to stay in business, is pretty stable in terms of the rate of change, is a genuine overhead, and should be treated as such in terms of ruthless cost management. On the other hand, the value from using new business technology or BT that over a third of CIOs are aspiring to, is focused on individual parts of the business in doing what they do uniquely, but doing it far better. That's not part of an enterprise-wide cost recovery overhead model of funding. It's a directly attributable cost to a specific business activity—and that's where the elasticity of the cloud computing model kicks in.

The pressure for new projects comes from two directions. One is from the cost-cutting CFO. The other is from specific functions directly related to the need for intelligence, decision support, and building new online products and services to sell. One of the key advantages of cloud computing is not just that we can build and deploy new business applications rapidly and at low cost, it's that we can implement new revenue-generating business models by using situational business processes in the Cloud. The challenge for the CIO is to make sure that this happens in a coherent manner in the context of the entire enterprise. Taking on such a role with a direct hand in supporting innovative business models is the challenge that a third of the surveyed CIOs are currently grappling with. Interestingly, these CIOs are working in the most successful enterprises.

In business life today, employees and those external parties who do business with our enterprises are evaluating and using technology as a key part of their work, and their business units' successes. In response to such changes it's time to consider how to adapt the funding model of the last century to one more suitable for the coming decades, especially in light of unpredictable change in the economy.

Maryfran Johnson, Editor in Chief of *CIO* Magazine further sets the context for the challenge, "Cloud computing seems to have moved from an over-hyped industry buzzword to a serious topic worthy of attention for many enterprise CIOs. There are still big unanswered questions hovering around security and integration issues with cloud computing. But the global economic recession is clearly accelerating CIO interest levels in alternative ways to deliver software and services to organizations that are demanding ever-lower IT expenses while clamoring for ever-higher levels of computing support for collaboration and customer service."

It's in these contexts that organizations need to make thorough assessments and then plan their strategies for adopting cloud computing—and leadership and guidance from the next-generation CIO is critical to making these strategic decisions.

Takeaway

With Business Technology (BT) the focus is on people, communications, and collaboration, not computers and data. The CIO role has evolved from custodian of the infrastructure under the CFO to a business leader with a seat at the executive table, focusing on people, communications, and collaboration. The next-generation CIO is a strategic agent for business transformation. In the most advanced firms the CIO is the most likely executive to move to CEO. The next-generation CIO is best positioned to manage the creative-destruction power of technology in the face of today's unexpected change and to craft corporate strategy.

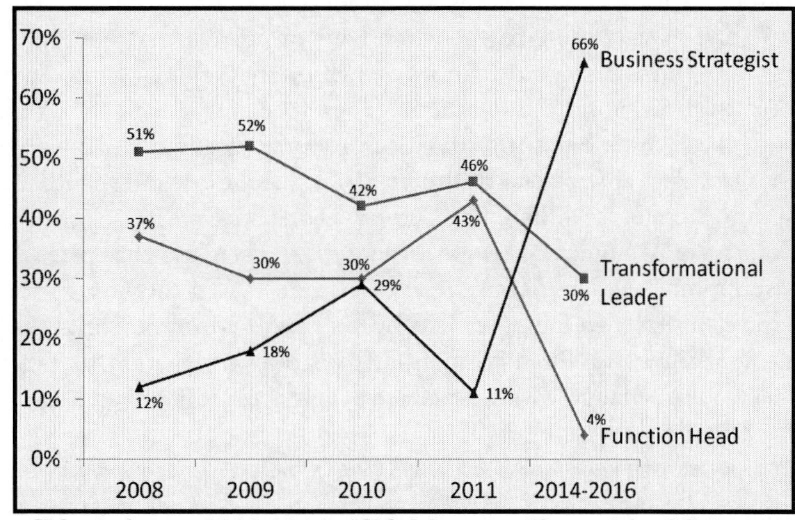

CIO Archetypes 2008-2016 (CIO Magazine "State of the CIO" 2011)

18. The Fractal Company

The Current State of Business

As globalization continues its march to becoming the greatest reorganization of the world since the Industrial Revolution, companies in a variety of industries are increasingly adopting a *choreography* role. They are building complex business ecosystems comprised of ever-growing numbers of highly specialized business partners and *prosumers* that place the consumer directly in the heart of business processes that produce value. *Back office* systems of record and transaction processing have been mastered, but *front office* systems for actually *conducting* business with disparate *systems of systems* that interact across the ecosystem are today's game.

A company is a complex social system, a system that must be treated as such, a complex organism of many dimensions and components – customers, production, and suppliers (and their information systems) fused together as One to create Value Networks. The days of the vertically-integrated, one-dimensional company are over. Linear Porter value chains have morphed into complex, multidimensional business ecosystems that have a lifecycle that goes from birth, to optimization, and on to a state of entropy as do all systems in the universe. Yet we still tend to model our enterprises and processes on a *flat plane* as rather static entities. Worse, Enterprise Architecture is often separate from BPM in the minds of many practitioners in today's companies.

Consider a snippet from Wolf Rivkin's book, *The Eloquent Enterprise,* "*Enterprise Growth Dialectics*: An Enterprise changes according to the same laws of change that govern any structure: the negation of the negation and the transformation of quantitative changes into qualitative ones according to the pattern of thesis-antithesis-synthesis. Therefore, any prospective framework describing this new qualitative state of an enterprise must offer a new, qualitatively different set of artifacts to describe the highest level of abstraction of the previous state. Only internal dialectic factors push an Enterprise from one Architectural State to another." Thus an Enterprise Architecture is a living, changing thing. "Enterprise Architecture is not a theoretical discipline that provides a nice-to-have variably detailed description of a static Enterprise or of one that is moving forward without deep structural changes and can therefore be described by a static, rigid framework. It is, rather, a must-have practical guide for Enterprise Transformation, providing a dynamic description of the profound but smooth structural change of an Enterprise that seeks effectiveness, efficiency, and the resulting competitive edge."

With the emergence of *Cloud-based, long-lived, loosely-coupled, stateless architectures,* existing Enterprise Architecture approaches and today's BPM systems will quickly demonstrate their lack of flexibility and inability to choreograph participants in complex business ecosystems made up of large numbers of actors spread across the globe.

To elaborate, let's turn to social visionary, Dee Hock, the founder and former CEO of VISA International, an organization that he says was founded on chaordic principles – a blending of chaos and order. VISA now connects over 20,000 financial institutions, 14 million merchants, and 600 million consumers in 220 countries. Hock explains, "By chaord, I mean any self-organizing, self governing, adaptive, nonlinear, complex organism, organization, community or system, whether physical, biological, or social, the behavior of which harmoniously blends characteristics of both chaos and order." In his view, today's current forms of organization are almost universally based on compelled behavior, or tyranny. The chaordic organizations of the future will embody community, based on shared purpose. A chaordic organization harmoniously blends characteristics of competition and cooperation [coopetition]. Hock's book, *Birth of the Chaordic Age,* hits broadside against the dominance of today's command-and-control institutions.

Hmm... Is the current state of the process management lifecycle based on "command-and-control?" Are current business modeling techniques ready to model a "self-organizing, self governing, adaptive, nonlinear, complex organism, organization, community or system, whether physical or social, the behavior of which harmoniously blends characteristics of both chaos and order?" Is it time to question the first principles of Enterprise Architecture and BPM in light of global Chaords that are neither centralized nor anarchical business networks?

Fractal Enterprise Architecture

Euclidean geometry is modeled by our notion of a "flat plane." Other geometries have been introduced to go beyond a flat plane, e.g., Elliptic and Hyperbolic Geometry. To illustrate, on a sphere, the sum of the angles of a triangle is not equal to 180°.

The very physics of traditional business processes might be ripe for change for the complex global enterprise goes well beyond a flat plane, and fractal geometry principles just may provide a fresh foundation for business modeling and Enterprise Architecture.

The main characteristic of fractals is *self-similarity*, implying *recursion*, pattern-inside-of-pattern. The term fractal was coined by Benoît Mandelbrot in 1975 and was derived from the Latin fractus meaning *broken* or *fractured*. A mathematical fractal is based on an equation that undergoes iteration, a form of feedback based on *recursion*. Mandelbrot sets display self-similarity because they not only produce detail at finer scales, but also produce details with certain constant proportions or ratios, though they are not identical.

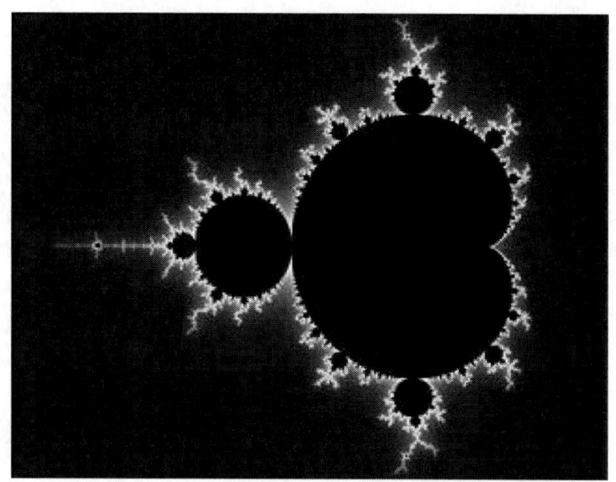

The Fractal Company Looks More Like This and Less Like a Hierarchal Org Chart

Because they appear similar at all levels of magnification, fractals are often considered to be infinitely complex. Natural objects that are approximated by fractals to a degree include clouds, mountain ranges, lightning bolts, coastlines, snowflakes, and thunderstorms.

In nature, small parts resemble the whole: The veins in leaves look like branches; branches look like miniature trees; rocks look like miniature mountains. That's kind of like a company's real business processes distributed throughout today's complex, global ecosystem comprised of ever growing numbers of business partners and social networks—the fractal business process.

We should ask, "Can complex, global 21st century enterprises be adequately described without embracing the concept of fractals?"

The fractal factory, the fractal company, the fractal business process? Not isolated to academia, Mandelbrot demonstrated the practical application of fractal geometry to financial markets in his 2006 book, *The (Mis)behavior of Markets: A Fractal View of Financial Turbulence.* See also:

tinyurl.com/7ctrne4

tinyurl.com/6ejvj4

Other practical sightings of fractals in business can be found in the emerging world of Agile Manufacturing, which is seen as the next step after LEAN in the evolution of production methodology. Let's turn to *A Comparison of Emerging Manufacturing Concepts* by Tharumarajah, Wells, and Nemes, "The concept of fractal factories proposes a manufacturing company to be composed of small components, or fractal entities. These entities can be described by specific internal features of the fractals.

The first feature is self-organization that implies freedom for the fractals in organizing and executing tasks. They may choose their own methods of problem solving including self-optimization that takes care of process improvements.

The second feature is dynamics where the fractals can adapt to influences from the environment without any formal hindrance of organization structure.

The third feature is self-similarity interpreted as similarity of goals among the fractals to conform to the objectives in each unit.

In addition to the above characteristics, there is a need for the factory fractals to function as a coherent whole. *This is achieved through a process of participation and coordination among the fractals, supported by an inheritance mechanism to ensure consistency of the goals.*

Fractals are structured bottom-up, building fractals of a higher order. Units at a higher level assume only those responsibilities in the process that cannot be fulfilled in the lower order fractals. This principle guarantees teamwork among the fractals and also forces distribution of power and ability in order to coordinate the actions of the individual fractals and put in place mechanisms that permit self-organization and dynamic restructuring."

If we consider the three main characteristics in relation to a fractal company, we can conclude that fractal architecture is fundamental to a business that is, after, all a complex adaptive system. In such an adaptive system, lower-level business fractals can influence other and higher-level fractals when new best practices are uncovered.

For example when Proctor and Gamble's Mexican operations got close to their lower-income customers, P&G learned that lower-income Mexican women take laundry very, very seriously. They cannot afford to buy many new clothes, but they take great pride in ensuring that their family is turned out well. P&G found that Mexican women spend more time on laundry than on the rest of their housework combined. More than 90% use fabric softener, even women who do their laundry by hand. That's no problem if all this is just a matter of pressing a button every once in a while. But it's no joke if you are doing the wash by hand or have to walk half a mile to get water.

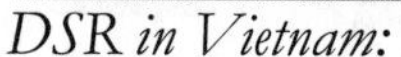

DSR in Vietnam: *tinyurl.com/7p2s6oo*

The big *aha* was discovering how valuable water was to lower-income Mexicans. Putting it all together, P&G knew that Mexican women liked to use softener; they had high standards for performance; and doing the laundry was arduous and time consuming, and required large amounts of water.

Having identified a problem (making laundry easier and less water-intensive), P&G turned to the labs for an answer. Their solution: Downy Single Rinse. Instead of a six-step process, DSR reduced it to three - wash, add softener, rinse saving enormous time, effort, and water. Launched in 2004, DSR was a hit from the start.

But, as fractals (P&G Mexico) influence other fractals throughout the P&G system, DSR became a huge branding success—worldwide.

Of course, other regions adopted localized messaging to take advantage of DSR (the self-similarity of fractals sharing a common goal).

By rinsing only once, the savings from your water bill covers the cost of DSR. So the message became, "Downy for Free!" from Mexico to Vietnam and all countries in between.

The unanswered question becomes, "Is Enterprise Architecture, as practiced today, ready to take on the fundamentals of the fractal company?"

A fractal company doesn't mean in any way a *fractured* company with chaos distributed via totally autonomous, disjointed entities or fractals.

> Self-similar fractals are the essential components of
> a networked, autonomous, cooperative, self-organizing, self-managed, complex adaptive system
> in the multidimensional business ecosystem called the 21st-century enterprise.

Let's not forget the term we discussed in the chapter on OODA loops, "schwerpunkt," the German term meaning *organizational focus*—that's what ties fractals together, even as they are self-managed, complex adaptive systems in and of themselves, in the Decide and Act part of OODA, as well as the ongoing Observe and Orient as we saw in the P&G example.

> Just as it takes fractal geometry to model a cloud in the sky,
> it takes fractal architecture to model the innovative business in the Cloud.

Read more on fractals in business:  *tinyurl.com/7mhbszd*

Is There a Mobile Autonomous Intelligent Agent in the House?

Okay, this new world of fractal Enterprise Architecture may make sense, but the "If-Then-Else" automation built into typical IT systems, even with Business Rules Engines, simply isn't going to cut it to build the complex adaptive systems needed in today's complex business ecosystems.

We suggest that what's needed is Intelligent Agent technology and ontological modeling methods. These ideas aren't new, but today's complexity calls for us to revisit them.

So, let's go *way back*, back before the explosion of the World Wide Web, back before the Enterprise Service Bus (ESB) and Cloud Service Bus (CSB), back before SOA, back before Services superseded Objects, back before the term BPM superseded the term BPR. From way back then: "Agent-Oriented Business Engineering (Farhoodi and Fingar, 1997) aims to combine the design of agent technology with business engineering to build a new area of research for the prize of enterprise-level computing. This methodology has two models:

- *The agent-oriented lifecycle model.* This model addresses domain modeling for agent-orientation by providing an active modeling metaphor and better analysis models that enable re-use.
- *The Ontology-based domain models.* Ontology defines the basic concepts and entities that are assumed to exist in some area of interest and the relationships that hold among them (a formal specification for conceptualization). This is a critical initial step in producing dynamic business-based systems."

"There are three distinct approaches to business-domain modeling: Business Process Reengineering (BPR), Object Oriented Technology (OO), and Intelligent Agents (IA). All these approaches are model-based and offer different techniques for describing problem domains. The BPR methods involve process, organization, events, business rules, entities, and relationships. The Object-Oriented Technology (OO) methods involve classes, objects, attributes, associations, operations, events, inheritance, polymorphism, and categories. These are well suited for software engineering modeling and have potential for reuse.

"However, they are not inherently business oriented, and provide premature commitments to design and implementation strategies. Intelligent Agent Technology (IA) can be leveraged to enhance enterprise modeling as well as offering new techniques for developing intelligent applications and smart technical infrastructure services. An agent-oriented perspective allows us to develop rich and expressive models of the enterprise and provide a foundation for adaptive and reusable business software. The convergence of OO, IA, and BPR results in a significant breakthrough in building models of the enterprise that is capable of end-to-end integration of business analysis and software systems." See: *Competing for the Future with Intelligent Agents:*

tinyurl.com/7eo2b82

The Great Dance of Business

If we consider intelligent agents as fractals in a multidimensional business ecosystem, some of the key components of *multi-agent problem-solving* are essential. To achieve common goals, agents need coordination. Effective coordination requires cooperation, which, in turn, can be achieved through communication and organization.

The difference is between active, central control and adaptive coordination—Orchestra versus Ballet. In orchestration, the conductor tells everybody in the orchestra what to do in real-time and makes sure they all play in synchronization. The conductor is an active leader, corrects for anomalies in real-time, and can introduce new information only he or she has. Orchestration in information systems also has an equivalent, the orchestration engine.

In choreography (Ballet), the choreographer coordinates the *plan* but is not part of execution. Each participant "listens to the music" and is responsible for its own adaptive behavior. In information systems each software agent "listens for events" and is responsible for its own adaptive behavior.

Distributed Intelligent Agents

Orchestration defines a procedure, and Choreography defines a protocol. Military style "command-and-control" management is giving way to "connect-and-collaborate," and the reason is clear. In today's world of total global competition, no one company is in control. Thus, orchestration of the entire business ecosystem by a single conductor is of days gone by.

Choreography via the foundation of peer-to-peer multi-agent systems, with autonomous and mobile agents sharing *common goals*, is the future. Of course, no sane person would hand over control of enterprise transactions to an unsupervised network of software agents. To investigate, negotiate, design, implement, monitor, and maintain complex, long-running enterprise partnerships you need a system that not only supports cross-boundary processes natively but has the right governance and policy management model. Such a model isn't based on the flow of work through a diagram, but on a framework including object types and distributed policy management (Role, Person, Interaction, Entity, Activity, Rule, and so on).

Just consider a human interaction management system (see Chapter 12) as a peer-to-peer multi-agent system. While Web 2.0 usage of the Internet has opened up new frontiers of possible human communication channels, it has also opened Pandora's box of noise, chaos, and distraction. If we turn to human interaction management systems we find that an agent-based platform can bring management control to all the noise, and the goal-oriented organizational design (GOOD) method can bring about chaordic collaboration despite the onslaught of Web 2.0 communication chaos. (See humanedj.com) Now this doesn't mean that workflow and integration style process management or ERP systems or transaction management systems "go away." Instead it means that agent-oriented systems appropriately sit on top of the enterprise software stack and act as front-office choreographers, taking care of business in the brave new world of virtual Value Networks.

Takeaway

If, as you should, want to continue this investigation of Fractal Enterprise Architecture and Agent-Oriented process management, here are some questions to ask:

- Can Enterprise Architecture be modeled with fractal geometry just as complex nonlinear systems are modeled in nature?
- Can such Enterprise Architecture become *executable,* at least as a simulation of the real business?
- Because what an enterprise "does" is far more important than "what it is," can process management be fused with Enterprise Architecture and both be executed in real-time?
- When a part of a complex business system "dies" (entropy) is that the end, or does ectropy extend the march to order? How does this apply to a business or its parts located in the midst of a global business ecosystem?

We're sure if JFK was asked such questions when considering putting man on the moon in the 1960s, many skeptics would have shouted, "Impossible. Far too complex. Get real." But the times were ripe back then. And, just perhaps, the time is now ripe for Fractal Enterprise Architecture and Agent-Oriented process management.

19. Epilogue: The Audacity of Innovation

"The future ain't what it used to be."—Yogi Berra

No More Old Normal. Huge transformations have been going on as globalization resets the economic playing field, and the very game of economic competition. Karl Marx was wrong about communism's inevitability. Alan Greenspan was wrong about large companies' inclination to self-regulate based on reputation. Milton Friedman was wrong about the gyroscopic capabilities of the Invisible Hand of the market with regard to economics.

The global economic problem at the time of writing this book lies deeper than the current slowdown and transcends the business cycle. So much for a business-cycle recovery from the 2007 Great Recession. It's far more likely to be a global economic reset the likes of the reset brought on by the beginnings of the Industrial revolution.

We repeat Jeremy Rifkin's observation, "The pivotal turning points in human consciousness occur when new energy regimes converge with new communications revolutions, creating new economic eras." So, keep in mind that as society goes, so goes business and commerce.

We will not get out of the downward economic spiral by doing more of the same things we've done before. We need new ideas, new business models. In short, we are going to have to innovate our way out of the mess and reinvent our future.

Let's recap five fundamental and interdependent shifts that are already affecting the very nature of a business:

1. The first shift stems from a monumental transition in the power balance between seller and buyer. To management's astonishment, the buyer is now in the driver's seat. As a result, the firm's goal has to shift to one of delighting customers—a shift from inside-out ("You take what we make") to outside-in ("We seek to understand your problems and will surprise you by solving them").
2. The second shift stems from the first transition, as well as the epochal transition from semi-skilled labor to knowledge work. Again to management's astonishment, traditional hierarchy suddenly doesn't work anymore. The role of the manager has to shift from being a controller to an enabler, so as to liberate the energies and talents of those doing the work and remove impediments that are getting in the way of work.

To support and sustain the first two shifts, the remaining three other shifts are necessary:

3. The mode of coordination shifts from hierarchical bureaucracy to dynamic linking, i.e. to a way of dynamically linking self-driven knowledge work to the changing requirements of clients.
4. There is a shift from value to values; i.e. a shift from a single-minded focus on economic value

and maximizing efficiency to instilling the values that will create innovation and growth for the organization over the long term. To cite Peter Drucker, "There is nothing so useless as doing efficiently that which should not be done at all. Management is doing things right; leadership is doing the right things."

5. Communications shift from *command* to *conversation*, a shift from top-down communications comprising predominantly hierarchical directives to communications made up largely of adult-to-adult conversations that solve problems and generate new insights.

None of these shifts is new. Each shift has been pursued individually in some organizations for some years. However when one of these shifts is pursued on its own, without the others, it tends to be unsustainable because it conflicts with the goals, attitudes and practices of traditional management. The five shifts are interdependent. When the five shifts are accounted for simultaneously, the result is sustainable change that is radically more productive for the organization, more congenial to innovation, and more satisfying both for those doing the work and those for whom the work is done.

Back to the future. A new economic system will evolve to match the global hyper-connected system that is itself emerging. We don't know how to evaluate this kind of tiny networked business by using the machine rules approach. But we did know how to do this in the past when we had local, instead of global, economies.

In 1900 most business was local. They did not have the Web to network but network they did. They used face-to-face networking, locally. Businesses developed extensive personal relationships not just with each person, but participated in the health of local business networks as well. Like a good gardener, they helped their local community get stronger. We used to know how to do all this. But of course it does not scale if we use a machine/industrial model.

To meet the opportunity of the new market and distributed, democratic capitalism, we also have to discover how to scale the older networking process. How can we possibly do this? We know that all natural systems are fractal. If we can design a tiny model based on network principles, we can have confidence that we can scale it.

Enter the Cloud—and business innovation in the Cloud!

Business Innovation is an organizational capability to do three things: 1) find new ideas, 2) convert them into value, and 3) distribute them throughout the organization so that the value can be achieved over and over.

> An innovation capability is a change maker,
> a disruptor for any organization.
>
> It's also natural.

We are all innovators by nature. In organizations, innovation can be developed into a *practice* with skills that are honed by the practitioners over time.

But that means "Doing the Work" of innovation, a repeating theme throughout this book. Once again, business innovation is all about execution!

We've covered a lot of ground, and now it's time to close with some of those proverbial "action items." What to do, what to do?

In the first half of the book we explored the current state of business innovation. In the second half we looked at multiple perspectives and key variables or ingredients to be included in the work of business innovation:

- OODA Loops as the Basis for Business Agility
- Open Innovation
- Social Networks as a Source for Business Insights
- Collective Intelligence and the Deliberatorium
- Explicit Collaboration
- Process Execution in the Intercloud
- Business Process Innovation
- Innovation as a Business Process
- Innovation as a Team Sport and RACI Matrices
- Big Data and Predictive Analytics
- The End of Management
- Open Leadership
- Services Innovation
- Persona Management Systems
- TRIZ
- Calling on IT Services Organizations
- Cloud Brokers
- The Next-Generation CIO
- The Fractal Company
- Business Architecture for the Innovation Economy
- Energy Maneuverability

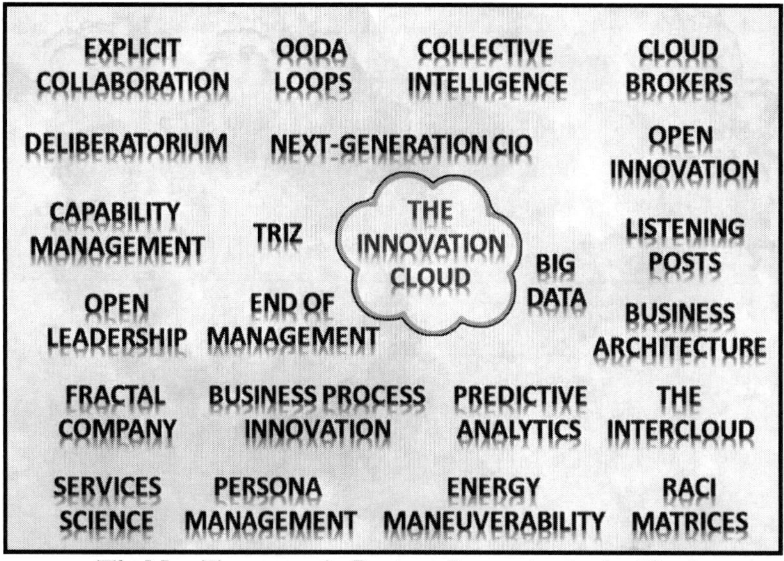

The New Taxonomy for Business Innovation in the Cloud

Get your house in order. Without going into the technical details, companies must embrace a new "business architecture" enabled by cloud computing: a Cloud-Oriented Business Architecture (COBA) rooted in the agility of OODA and the foundation of business agility embodied in Boyd's energy maneuverability analysis.

tinyurl.com/7ljcd4u

As shown below, the foundation for a refreshed business architecture rests on existing, legacy assets and current business operations which are in turn supported with current information technology. These assets and modes of operation aren't going to be torn up and discarded any more than the Interstate highway will be torn up and discarded in any new era of transportation. They are the current lifeblood of any organization. They will, however, be integrated into the new house of business via service-oriented architecture.

There are six new pillars organizations will need to place atop this foundation in order to have the structures needed in the Cloud Economy: Social Networks, Demand-Supply Chain Business Networks, Distributed Business Rules Management, Next-Generation Analytics, Capability Management and Innovation Management.

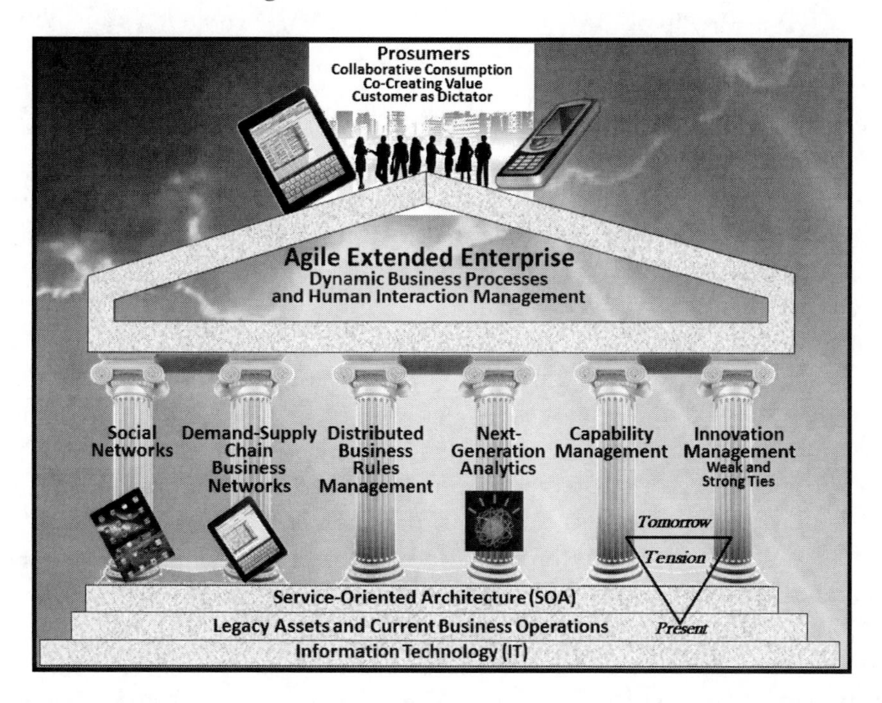

Capability Management is of particular concern as a means of balancing the current way of thinking and operating with tomorrow's way of thinking and operating. Here's why. In *The Other Side of Innovation*, professors Vijay Govindarajan and Chris Trimble note that there are two kinds of teams in most firms: 1) the Performance Engine, which is the portion of the business focused on the day to day execution of the business and focused on earning profits, doing things consistently and efficiently, and 2) a Dedicated Innovation Team which is typically in direct conflict with the Performance Engine.

The Innovation Team talks about "breaking all the rules" which "sounds like breaking the Performance Engine." There is direct conflict between the goals and expectations of the two teams.

To achieve breakthrough innovation, the Dedicated Innovation Team must act as though it's building a new company from scratch. For the Performance Engine to do that would require forgetting and unlearning, two nearly impossible tasks for humans. Anyone who has created a new project and attempted significant change within an ongoing organization knows about this conflict. Hence, a strong focus on capability management is needed to manage the tension between the present and tomorrow, between the as-is and the to-be.

The Innovation Team must leverage the capabilities of the Performance Engine, or else it would be stripped down to the resource level of a garage startup. Capability management is key to forming a partnership between the two teams. Pioneered by the military, capability management aims to balance current operational requirements, with the sustainable use of current capabilities, and the development of future capabilities, to meet the sometimes competing strategic and current operational objectives of an enterprise. Accordingly, effective capability management:

- Assists organizations to better understand, and effectively integrate, re-align and apply the total enterprise ability or capacity to achieve strategic and current operational objectives; and
- Develops and provides innovative solutions that focus on the holistic management of the defined array of interlinking functions and activities in the enterprise's strategic and current operational contexts.

While a *capability* represents the "what" that a company needs to execute its strategy, a *business process* represents "how" the capability is brought to life. It's also useful to distinguish between *Sustaining Innovation* for the Performance Engine versus *Disruptive Innovation* as it relates to the Dedicated Innovation Team. As we discussed in Chapter 4, both types of innovation are needed by the Agile Extended Enterprise to meet the challenges of 21st century competition. While Disruptive Innovation may be the homerun, Breakthrough Innovations are the singles and doubles that count when Innovation and Performance are truly working and learning together.

These six pillars are needed to build the Agile Extended Enterprise where 20th century IT centered on *productivity* is superseded by 21st century business technology (BT) centered on *collaboration* at both the human and business process levels. This new house of business will serve prosumers that directly participate in the co-creation of value—and where the customer as king gives way to the customer as dictator.

While it is overly simplistic to claim that all organizations are dealing with the same obstacles to innovation, the fact is that there are repeating themes and patterns of obstacles that will need to be addressed if you expect to establish a sustainable culture of innovation:

- Lack of a shared vision, purpose or strategy
- Mistaking innovation for R&D and not balancing *exploration* (R&D) with *exploitation* (innovation)
- Intolerance for ambiguity
- Innovation not articulated as a company-wide commitment nor on goal sheets. One team re-

sponsible or everyone's job? (centralized versus decentralized)

- Lack of committed leadership, making innovation accidental instead of planned
- Constantly shifting priorities
- Internal process focus rather than external customer-outcome focus
- Focus on successes of the past and what is working now rather than the challenges of the future
- Politics and efforts to sustain the status quo to support entrenched interests; open versus closed innovation
- Rewarding crisis management rather than crisis prevention
- Command-and-control management of new ideas*: micromanagement*
- Short-term thinking and under-funding of new ideas in the name of sustaining current efforts and unable to balance the radical innovations needed for the long haul versus the incremental innovations needed day to day
- Reluctance to kill initiatives that are not succeeding, but have been funded and staffed
- Fear that criticizing current practices and commitments is a high-risk activity
- Workforce workloads: too much to do, not enough time
- Risk aversion and punishment for "failure"
- Lack of useful systems and processes for innovation and wishful thinking that innovation will simply emerge
- Convergent thinking at the cost of divergent thinking; unable to balance analysis with intuition
- Unwillingness to acknowledge and learn from past "failures"
- Inadequate understanding of customers and self in order to neutralize (bring up weaknesses of competitors) or differentiate (extend core strengths)
- Lack of time and training to pursue creative thinking or to develop new ideas and opportunities

Let's ponder some thoughts from the world of science instead of the world of business. Michael Nielsen's book, *Reinventing Discovery: The New Era of Networked Science,* discusses how our improved ability to network via the Internet is changing the face of science. One of the stories that he tells is of a tournament sponsored by playchess.com in 2005. It allowed humans and computers to enter together as hybrid teams. The going-in favorites for the tournament were the teams based upon the Hydra chess playing computer, who had dominated for years.

It didn't turn out that way, "The grandmasters could beat the Hydras because they knew when to rely on their computers, and when to rely on their own judgment. Even more interesting, the winner of the tournament was a team called ZackS that consisted of two low-ranked amateur players, using three off-the-shelf computers, and standard chess-playing software. Not only did they outclass the Hydras, they outclassed the grandmasters with their strong chess-playing computers. The human operators of ZackS demonstrated exquisite skill in using the data-driven intelligence of their computer algorithms to amplify their chess-playing ability.

"As one of the observers of the tournament, Garry Kasparov, later remarked, "Weak-human + machine + better process was superior to a strong computer alone and, more remarkably, superior to a strong human + machine + inferior process."

We can only ponder the implication of this weak-human + machine + better process for the game of business.

Let No Collapse Go to Waste

A Study of History is the 12-volume magnum opus of British historian Arnold J. Toynbee, finished in 1961, in which the author traces the development and decay of all of the major world civilizations in the historical record. Toynbee applies his model to each of these civilizations, detailing the stages through which they all pass: genesis, growth, time of troubles, universal state, and disintegration. Here's Toynbee's basic thesis, taken from Wikipedia, "He argues that the breakdown of civilizations is not caused by loss of control over the environment, over the human environment, or attacks from outside. Rather, it comes from the deterioration of the "Creative Minority," which eventually ceases to be creative and degenerates into merely a "Dominant Minority" (who forces the majority to obey without meriting obedience). He argues that creative minorities deteriorate due to a worship of their "former self," by which they become prideful, and fail to adequately address the next challenge they face." Can we substitute the word "business" for the word "civilization" here?

Reread the quote and decide for yourself.

In his 2005 book, *Collapse: How Societies Choose to Fail or Succeed,* Jared M. Diamond, professor of geography and physiology at University of California agrees with Toynbee, that "civilizations die from suicide, not by murder" when they fail to meet the challenges of their times. Diamond identifies five factors that contribute to collapse: climate change, hostile neighbors, collapse of essential trading partners, environmental problems, and failure to adapt to environmental issues.

Now that the U.S. Supreme Court has declared that corporations are people, and we know that people make up civilizations, companies should pay close attention to Toynbee and Diamond for it's our very civilizations that are now experiencing collapse. The daily news is full of shocking stories on all the fronts discussed by Toynbee and Diamond.

So how do we get out of this apparent mess? Is there anything we can do to avoid collapse? Is there any reason for hope in these turbulent times? In their book *Abundance: The Future Is Better Than You Think,* Peter H. Diamandis, the CEO of the X-Prize Foundation, and science writer Steven Kotler argue that new technologies and other forces will make the coming years better, not worse, as we so often worry. Their main point is that when seen through the lens of technology, few resources are truly scarce; they're mainly inaccessible. Yet the threat of scarcity still dominates our world view.

tinyurl.com/867vo4z

Diamandis and Kotler explore how four emerging forces—exponential technologies, the DIY innovator, the Technophilanthropist, and the Rising Billion—are conspiring to solve our biggest problems: water, food, energy, healthcare, education and freedom. They document how progress in artificial intelligence, robotics, infinite computing (the Cloud), ubiquitous broadband networks, digital manufacturing, nanomaterials, synthetic biology, and many other exponentially growing technologies will enable us to make greater gains in the next two decades than we have in the previous two hundred years. We will soon have the ability to meet and exceed the basic needs of every man, woman, and child on the planet. Abundance for all is within our grasp.

Before we close this book let's turn to Dutch futurist, Adjiedj Bakas, author of *Beyond the Crisis: The Future of Capitalism.* Bakas writes, "The Chinese phrase for crisis consists of two words danger and opportunity. The current economic crisis marks the end of an era, and the start of a new one. During the current crisis we finally say goodbye to the 20th century and transit into the new, post-industrial economy with new economic pillars and the mixture with digital collaboration powered by 'cloud computing,' the next generation Internet. This transition period offers tremendous opportunities, for out of the current crises, a new economy will emerge."

www.mkpress.com/beyond

Taking on the work of *cloudy* business transformation is scary stuff to say the least, for we have no prior frame of reference. So, take the advice of David Allen, author of *Getting Things Done*, "Our productivity is directly proportional to our ability to relax. Only when our minds are clear and our thoughts are organized can we achieve effective productivity and unleash our creative potential." So, "relax" and get on with doing the audacious work of business innovation in the Cloud. Your business, your family and the whole world—they are all counting on you!

Let no collapse go to waste. Innovate or die.

Carpe diem!

Appendix A: Bibliography

This online bibliography contains 5 content sections:

- Innovation
- Social Media
- Web 2.0
- Cloud Computing
- Service Orientation
- Business Process Management (BPM)

Suggested Readings are indicated with an asterisk (*).

www.mkpress.com/CloudReading/
or tinyurl.com/7hhrtya

Please send your suggestions for updating the bibliography to
susan@mkpress.com

Index

1

1984, Orwell 156

3

3D printing 74

4

4Inno.com 131

A

A Guide to Open Innovation and CrowdSourcing 143
Accenture 100
accountability 230
Act 97
Adaptive Case Management 238
affective computing 149
affective ontologies 151
Agent-Oriented Business Engineering 272
Agile Manufacturing 270
agility 85, 102
Air Force Research Laboratory 61
Allen, David 285
AlloSphere 68
alphabet as a technology 221
Amabile, Teresa 88
Amara's law 24
Amazon 77, 78
ambient intelligence 206, 214
amorphous networks of peers 221
analytics 145, 208
Anderson, Chris 256
Apple 114, 146
applications 53
Architecting the Internet of Things 214
architecture 47, 116, 121
argumentation system 162
ARM processors 57
Arthur, Brian 64
artificial intelligence 156
artificial neural network 57
artificial photosynthesis 105
Atala, Anthony 75
atomic data storage 215
attention allocation metrics 160
Attention Mediation: 162
auditability 54
augmented reality 72
authentication 55
autonomous decision-making 214
autonomous logistics objects 215
autonomous, self-managed teams 228, 230
autopoiesis 65

B

Bakas, Adjiedj 285
Barnatt, Christopher 75
Bartlett, Albert 38
battery systems 106
Bayesian inference 214
Benioff, Marc 226
Benyus, Janine 232
Beyond the Crisis: The Future of Capitalism 285
Big Data 67, 150, 207, 211, 212, 252
big-data analytics 25
biomimicry 232
bioteaming 232
Bioteams 222
Birth of the Chaordic Age 268
Bonk, Curtis 232
Booz Allen Hamilton 88
Booz and Company 71, 171
Boston Consulting Group 34
Boyd, John 96

BPM as a Service (BPMaaS) 54, 111
Breakthrough Innovation..89
Bricklin, Dan 28
Bring Your Own Device (BYOD).........................46
Brown, John Seely 114
Bushell, Sue...262
business innovation.......................... 28, 69, 70, 277
business intelligence (BI)....................187, 189, 210
business mashups 29, 188
business model innovation....................................89
Business Networks 53, 222
business operating system 54
Business Operations Platform (BOP) 54
business process... 192, 236
business process innovation 101
Business Process Management (BPM) 54
Business Process Management System (BPMS) . 54
Business Process Management: The Third Wave........191, 234, 236
Business Process Outsourcing............................251
Business Services as a Service............................251
business services Cloud 27
Business Singularity...222
business technology (BT) 262
Business Week.. 33, 114
business-savvy CIO...262

C

Capability Management 280
CAPEX 50
CapitalCommunism...82
Capture ideas 36, 173
case management..198
Caudell,Tom 73
CCNx project 209
Cemex..241
CENCOR 29, 112
Cerf, Vint 249
Champion 37, 173
change maker...277
chaords...268
Charan, Ram 136
Chicago Strategy Associates 187
Chief Cloud Officer..263
Chief Polymath Officer..134
Chief Process Officer 262
China 69
Chomsky, Noam 145
choreography 53, 54, 167, 203, 238, 267, 273
Christensen, Clayton...114
CIO 260, 262, 264
CIO Archetypes 266
CIO Magazine 262, 266
Cisco..187, 206
Clarke, Arthur.. 22
closed shops 196
Cloud *appliances* 250
Cloud Brokers.............................168, 251, 254, 255
cloud computing 49, 50, 51
Cloud Enterprise 221
Cloud Security Alliance 55
Cloud service providers (CSPs) 127, 252
Cloud services 53
cloud sourcing 34, 241
Cloud Standards Customer Council............55, 256
Cloud, the 24, 51
Cloud-Oriented Business Architecture (COBA) 279
clusters.. 69
co-create value 164, 177
co-creation of innovation 128
cognitive computer 56, 58
cognitive science models..................................... 203
collaboration 26, 135, 189, 261, 281
Collaborative Engineering system...................... 111
collective intelligence.......................26, 34, 158, 181
Collective Leadership 233
command-and-control................................221, 231
CommuCapitalism 82
Competing on Analytics.. 210

complex adaptive systems 227
connect-and-collaborate 222, 231
Consortium for Service Innovation 244
constraints 119
Content-Centric Networks 19, 209
context 68
coopetition 24, 188, 268
co-production 237
Cordys 111
Corporate Executive Board 257
creative destruction 71
Creative Destruction 2.0 78
Creative Destruction 3.0 82
Creative Thinking 88
creativity 88, 119
crisis-driven choice 41
critical thinking 154, 160
Cronkite, Walter 217
cross-boundary processes 274
crowd sourcing 26, 128
Culture of Innovation 171
Customer Age 187
customer insights 187
customer-driven 30
customer-facing processes 187
customers 189
Cutting, Doug 209
cycle-time management 102

D

Darwin, Charles 86
Davenport, Tom 211
Decide 97
decision cycle 96
decision markets 151
decision models 210
Declaration of Interdependence 229
Dedicated Innovation Team 280
Defense Advanced Research Project Agency (DARPA) 57
deliberation map 160
Dell 234
Dell's Ground Control 148
Dell's IdeaStorm 129
Delphi method 181
demand-supply value chain 166
Deming, Edwards 38, 112, 261
Design by Committee 196
Design for Six Sigma 112
Develop and experiment 37, 173
diffusion of innovation 69
digital Balkanization 196, 217
Digital Natives 44
Digital Rights Management 55
direct digital manufacturing (DDM) 74
Disruptive Innovation 89
Do the Work 32
Doblin 113, 114, 124
doing the right thing 21
doing things right 21
Dot.Cloud: The 21st Century Business Platform) 165
Downy Single Rinse 271
drones 154
Drucker, Peter 28, 68, 85, 171, 277
Dynabook 23

E

Economics 2.0 19
ecosystem versus *organism* 68
ectropy 77
Edison, Thomas 25, 70
EDS 234
Embedded Systems Institute 122
emergent processes 192
emotions 149
employee-owned assets 46
Employees first, Customer's Second 231
encapsulation 196
End of Management 227
end-to-end business processes 54, 167, 177

Energy-Maneuverability theory 96
Enomaly 255
Enterprise 2.0 18
Enterprise Architecture (EA) 117, 268, 271
Enterprise Cloud Computing 53, 264
Enterprise Transformation 267
entropy 77
Evaluate and select 36, 173
Everything as a Service 49, 51, 68, 261
exabytes 66, 151, 207, 215
execution 37
executive compensation 229
experiences 236
Expertise (Knowledge) 88
explicit collaboration 197
exponential change 38
eXtensible Access Control Markup Language (XACML) 55
Extreme Competition 70

F

F-16 96
fail fast 37
failure 21, 29
Fast Company 98, 101
fast transients 97
fear, uncertainty and doubt 54
FedEx 115
feelings, not *facts* 150
filtering algorithms 217
First Digital Revolution 76
flat plane 267, 268
Forbes 208
Ford 242
Ford, Henry 63
Foresight 118, 212
Forrester Research 200, 225
fractal 268
fractal company 271
Fractal Enterprise Architecture 103
fractal geometry 269, 272
Friedman, Milton 85, 276
Furber, Steve 58

G

Gall, John 25
game changer 52
Gartner 79, 116, 173, 207, 254, 256
gatekeepers 217
Gates, Bill 28
Gatorade Mission Control Center 147
Gause's law 24
GE Global Research 28, 234
GE's Ecomagination 128
Gelernter, David 218
General Electric 33, 245
General Motors EN-V 242
genetic algorithms 213
geolocation 153
Gerstner, Lou 234
Getting Things Done 285
Ghalimi, Ismael 201
gift economy 172
GigaSpace Technologies 50
Gilding, Paul 40
Gladwell, Malcom 84
Global Footprint Network 39
globalization 27
goal-oriented organizational design (GOOD) .. 274
Goldcorp 128
Goodhart's law 27
Google 249
Google Analytics 252
Google's driverless car 243
Govindarajan, Vijay 280
GPS 72
Granovetter, Mark 135
Gray, Dave 241
Greenspan, Alan 276

H

Haas, Harald 215
Hadoop 208, 209
Halal, William 181
Hamel, Gary 18, 222
Hammer, Michael 20
Hammonds, Keith 98
happiness-driven growth model 40
Haque, Umair 81
Harrison-Broninski, Keith 201, 238
Harvard Business Review 102
Hawken, Paul 81
HCL Technologies 230
Heyak, Fredrick 82
hierarchies 226
history of management 222
Hock, Dee 268
Holland, John 228
Hollywood business model 43
HP 78, 135, 152, 234
Hugos, Michael 201
Human Interaction Management 89, 194, 198
human-to-human interactions 192, 238
Hydrocarbon Man 106
hypothesis-to-experiment cycle time 213

I

IBM 234, 235
IBM's Global InnovationJams 133
identity management 55
IDEO 115
IIT Institute of Design, ... 124
Immelt, Jeffrey 245
Implement 37, 173
implicit collaboration 194
in-box hell 199
India 69
Industrial Age 187
infoglut 191
information Cloud 27
Information Technology 2.0 19
Information Technology Industry Council 244
information wants to be free 196
Infrastructure as a Service 49, 50, 110, 254
InnoCentive 131
innovate to zero 48
innovating innovation 114
innovation 27, 69, 88, 91, 114
Innovation Architecture 117, 147
innovation as an Episodic Event 29
innovation consumption 91
Innovation Economy 33
Innovation Group 177
Innovation Hype Curve 27
innovation in management 223
Innovation Landscape 113
Innovation Log and Feedback 185
Innovation Plan 184
innovation platforms 138
innovation process lifecycle 175
Innovation Process Management (IPM) 29, 112
innovation processe steps 173
Innovation Roadmap 184
In-Q-Tel 149
inside-out 28
Insight 119, 212
Institute for Business Value 244
Intel 161
intellectual property (IP). 127
intelligent agent technology 215
Intelligent Agents 273
intelligent objects 214
Intercloud 35, 165, 169, 248
internal IT 246, 257, 259
Internet of Things 57, 207, 214
Intuit 239
Invisible Hand 276
invisible hand of transparency 222
iPhone 29
iPod 69, 89, 90

Irving Wladawsky-Berger ... 234
IT Consumerization ... 46
IT Doesn't Matter ... 52
IT services industry ... 247
iTunes ... 89, 90

J

Jacobson, Van ... 209
Jaruzelski, Barry ... 88
Jeopardy! ... 147, 209
jobless society ... 85
Jobs Steve ... 65
Johnson, Maryfran ... 266
Joseph Schumpeter ... 71

K

Kahn, Herman ... 179
Kapoor, Ajit ... 109
Kay, Alan ... 23
Keeley, Larry ... 114
Kelly, Kevin ... 64, 253
Kindle tablet ... 78
knowledge economy ... 33
knowledge worker ... 78
Korea ... 69
Korhonen, Janne ... 203
Koulopoulos, Tom ... 41
Krugman, Paul ... 151
Kuhn , Thomas ... 64
Kurzweil, Ray ... 63

L

Lafley, A.G. ... 130
Lanier, Jaron ... 195
Leadbeater, Charles ... 141
leadership ... 87, 222, 226
learning-based model ... 37
LED light bulbs ... 216
Levitt, Theodore ... 20
Lewis, Robert ... 104
Li, Charlene ... 225
Li-Fi (light fidelity) ... 216
listening posts ... 147
Lohr, Steve ... 156
long tail ... 159, 253, 257
Lytro camera ... 72

M

macro-level scenarios ... 181
mamagement, history of ... 35
manage without managers ... 228
management ... 30, 35
management as a technology ... 221
management control ... 54, 230
Management Innovation eXchange ... 18
Mandelbrot sets ... 268
Mandelbrot, Benoît ... 268
Manufacturing Consent ... 91, 145, 154
Marketing Myopia ... 20
Martin, Roger ... 81
Marx, Karl ... 276
mashups ... 53, 166
mass atomization ... 76
mass customization ... 165, 212
Mastering the Unpredictable ... 238
McCarthy, John ... 23
McGregor, Mark ... 69
McKinsey & Company ... 101
McKinsey Global Institute ... 206
McKinsey Quarterly ... 30
Mechanical Turk ... 163
Megatrends ... 17
Micro Air Vehicles ... 61
micro-innovation ... 114
Microsoft ... 114
Microsoft Research ... 60
middle class income ... 81
Millennials ... 44
Mind-Inside-the-Head [MITH] model ... 135
Mirchandani, Vinnie ... 134

Mirror Worlds 218
MIT Center for Collective Intelligence 158, 160
Mitra, Sramana 82
mobile Internet 49, 214
Mobile Processes based on Pi Calculus 202
modern economists 19
Modha, Dharmendra 57
Montgomery Ward 77
Morris, Robert 235, 244
motivation 88
MP3 player 89, 90
multi-agent problem-solving 273
multi-agent systems 203, 214
multi-company value chains 53
multi-path Internet protocol 249

N

Naisbitt , John 17
nanotechnology 57
Narrative Science 156
narrowcast 240
Natural Capitalism 81
Natural Language Processing (NLP) 150
Nature of Technology 64
Nayar, Vineet 231
Negative Income Tax (NIT), 85
negotiate-and-commit speech acts, 202
Nelsonin, Ted 209
netbooks 46
neuroscience 57
new IT stack 204
new meaning from new sources 208
New Normal 47
New York Times 156
Newton, Isaac 30
next practices 69
Ng, Ren 72
Nicholson, Geoffrey 28, 63
Nielsen, Michael 282
NineSigma 131
NIST Reference Architecture 250
no one in charge 161
Nocera, Daniel 105
noise 158
Ntrepid 153
Nussbaum, Bruce 33, 114

O

Object Management Group 55, 256
object-to-object communications 214
Observe 97
offensive strategy 96
OmniTouch 60
on-demand business innovation 53
on-demand computing 50
online persona 46
ontologies 214, 272
OODA Loop (Observe, Orient, Decide, Act), 97, 102, 109
Open Analytics 213
open editing 160
Open Group 251
open innovation 28, 127, 136
open source 28, 127
operational innovation 70
Operational Transformation 101
opinion mining 150
orchestration 167, 273
organic processes 192, 238
organizational innovation 70
Orient 97
Orwell, George 156
other points of view 217
outside-in 28

P

pace of innovation 29, 124
Palmisano, Sam 133, 209
Palo Alto Research Center (PARC) 70
Pariser, Eli 217

passion 222, 229
pay-as-you-go costing model 51
peer pressure 230
peer-to-peer knowledge webs 237
peer-to-peer multi-agent systems 274
PepsiCo 147
Performance Engine. 280
Periodic Innovation Reviews 184
Perot Systems 234
persona management systems 153
personal digital assistants (PDAs) 77
personal software agents 214
personalization 165, 217
person-to-object communications 214
pervasive Cloud 68, 248
pervasive simplification technology 56
Pink, Dan 18
Plan of Intent 121, 178, 184
Plan of Investigation 121
Plan of Record 121, 178, 182
Plan, Do, Check and Act (PDCA) 112
planning-based model 37
Platform-as-a-Service 49, 50, 110, 249, 254
Porter, Michael 81
Prahalad, C.K. 164
prediction markets 151, 181
predictive analytics 210, 211, 213
Presence 248
Pressfield, Steven 32
private information spaces 203
Process on Demand 238
Procter & Gamble 28, 33
Procter & Gamble 'Connect and Develop 131
product and service innovation 70
product invention 30
productivity 26, 261, 281
products as services 241
Prophet of Innovation 71
prosumers 239, 67, 281
proxemics 61
public opinion polls 151
punctuated equilibrium 48
PWC 173
Pyke, Jon 192

Q

QuickBooks 239

R

R&D labs 28, 78
RifkinJeremy 81
RACI (Responsible, Accountable, Consulted, and Informed) 189
RACI Matrix 190
radio frequency spectrum 215
RAND corporation 179
rapid elasticity 24
REACT/AIM 203
read-write-*execute* Internet 261
reality mining 212
Recorded Future 148
Reengineering the Corporation 20
reference architectures 121
Reich, Robert 82
Reinventing Discovery: The New Era of Networked Science 282
relevance 217
RFID tags 206, 212, 249
Rockefeller Foundation 181
role activity theory 202
Rosen, Ken 208
Rotman School of Management 81
Roubini, Nouriel 82
Russia 69

S

safety stock 236
Sakti3 106
Salesforce.com 226
Samsung 34, 146

sandbox 178
Santa Fe Institute 228
Sastry Ann Marie 106
scenario planning 178, 181
Schmidt, Eric 217, 249
Schumpeter, Joseph 30, 71
schwerpunkt 103, 104, 272
Second Digital Revolution 76
security 46, 54
Security Assertion Markup Language (SAML) 55
Seely Brown, John 48, 200, 239
self-service IT 49
self-similarity 268
sell-side innovation 70
Senge, Peter 245
sentic computing 151
SenticNet 150
sentiment analysis 150
serial planning mindset 37
Service Avatars 241
service blueprint, the 243
Service Design Network 244
Service Process Management (SPM) 238
Service Research and Innovation Institute 243
service science 235
service value chain 237
service-oriented architecture (SOA) 53, 201
services processes 236, 238
services sector 234
Services-as-a-Service 234
Shadow IT 249
Shannon, Claude 158
Shannon, Mark 158
shared vision 281
Shift Happens 33
Shill, Walt 100
Shirky Principle 109
Shirky, Clay 109
simple rules of engagement 172
singularity 63, 222
Siri 146
Situational Apps 54
situational business processes 53
Six Sigma 29
Smart Car 242
Smart Grid 242
smart phone 46, 73, 248
Smith & Hawken 81
Smith, Adam 221
Smith, Howard 114, 234
Social Business Council 244
social constructionism 145, 147
social Enterprise 262
social networking 26, 34
social networks 46, 127, 158, 214
Social Web 241
sockpuppets 153
Software as a Service 49, 50, 110, 254, 256
solar power 81, 105
SPARC 29, 113
Spence, Michael 82
Spoken Web 43
SpotCloud 255
Star Trek 24
STEEP Framework 118
Sterling Commerce 255
Stone Soup 126
Storage-as-a-Service 249
straight thru processes (STP) 201
Strangers to Ourselves 109
strategy 100
strong ties 135
Suggestion Box 230
supply-side innovation 70
surface computing 59
Surowiecki, James 152
sustainable innovation 116
Sustaining Innovation 89
Swann, Joseph 70
Swensen, Keith 238

SWOT (strengths, weakness, opportunities and threats)....121
Systemantics: How Systems Really Work and How They Fail....25
systemic culture of innovation....177
Systems of Neuromorphic Adaptive Plastic Scalable Electronics (SyNAPSE)....57
systems thinking....179

T

tablet computer....23, 46, 248
Taleb, Nicholas....82
tangibly....119
Tarde, Gabriel....91
Tata....106
team performance....230
TechCast forecasting system....181
technium....65
technological nationalism....69
technology....63
technology Cloud....27
Telxon....264
temporal analytics engine....148
tenets of innovation....86
The....152
The 2020 Workplace....44
The 33 Strategies of War....100
The Age of Customer Capitalism....81
the bullwhip effect....236
The Capitalist Manifesto....81
The Deliberatorium....160, 161
The Economist....50, 75, 127
The Eloquent Enterprise....267
The Engineering of Customer Services....234
The Fifth Discipline....245
The Filter Bubble....217
The Future of Corporate IT study....257
The Great Disruption....40
The Greatest Innovation Since the Assembly Line....201
The Lights in the Tunnel....84
The New Age of Innovation,....164
The New Polymath....134
The Next Convergence....82
The One....34
The Other Side of Innovation....280
The Real-Time Enterprise....101
The Second Digital Revolution....76
The Singularity Is Near....63
the social life of devices....206
The Structure of Scientific Revolutions....64
The Use of Knowledge in Society....82
Theory of Games and Economic Behavior....25
Theory X....229
Third Industrial Revolution....81
Thompson, Ken....142, 232
Three Laws of Prediction....22
Thrun, Sebastian....243
time-sharing....23
Toyota....29, 34
transparency....172, 230, 231
TRIAD project....209
TRIZ....115
trolls....148, 154
trust....171
Trust Management Systems....155

U

ubiquitous computing....214
Understand and Scope....36, 173
unexpected change....182
Unified Communications....194, 248
Unified Messaging....248
unintended consequences....90
units of business....53
units of technology....53
UPS....166
Uptime Institute....50
urbanization....253
utility computing....50

V

value 21
value chains 164
Value Challenges 184
Value Networks 267
Virgin Group 188
virtual desktops 51
virtual enterprise networks 177
Virtual Enterprise Networks 24
virtual worlds 76
VISA International 268
visible light spectrum 216
Visible Technologies 149
Visicalc 28
voice of the customer 188

W

W. L. Gore 226, 228
Wall Street 50
Walmart 77, 239, 264
watchlists 160
Watson 67, 209
Watson Research Lab 235
weak signals 212
weak ties 135
Web 2.0 .... 17, 189, 195, 214
Web 2.0 mob 197
Web services 53
Weisure Time 44
Welch, Jack 31
What Technology Wants 66
What's Mine is Yours: The Rise of Collaborative Consumption 81
white space opportunities 212
white-collar work 33
Whole Foods 229
Wikipedia 196
Will.i.am 3
Windstalk energy farms 108
Wired 256
wireless broadband 216
wireless data from every light bulb 215
wisdom of crowds 26, 34
Work Mobility 43
work-centered hierarchies 226
workflow 135, 238

X

Xerox 239

Y

Yet2.com 131
You Are Not A Gadget 195
Young, Jasmine 112
YourEncore 131

Z

Zipcars 68

About the Authors

JIM STIKELEATHER. For more than 30 years, Jim has designed, developed and implemented award winning information and communications technologies that help businesses and institutions succeed. He has spoken and consulted internationally on digital infrastructures, evaluation of emerging technologies, and provided strategic guidance on their application to achieve business outcomes. He participates in international technology standards bodies, has multiple book and industry-article contributions to his credit and advises a number of technology incubators. Additionally, Jim holds two patents. As Executive Strategist / Chief Innovation Officer for Dell Services, the approximately $8 billion IT services arm of Dell, Jim leads a team of information technology and business experts who identify, evaluate and assess the future potential of new technologies, business models and processes to address evolving business, economic and social trends for the company and customers. Jim came to Dell via Perot Systems. Perot acquired a company he helped found, the Technical Resource Connection (TRC), in 1996. He left in 2000 to become CTO of the New Ventures unit of MeadWestvaco, where he served as an executive and board member of their portfolio of startup companies. He returned to Perot Systems in 2009 to lead the development and execution of Perot's Cloud strategy.

PETER FINGAR, Executive Partner in the business strategy firm, Meghan-Kiffer Research, is one of the industry's noted experts on business process management, and a practitioner with over forty years of hands-on experience at the intersection of business and technology. As a former CIO and college professor, Peter is equally comfortable in the boardroom, the computer room or the classroom where he has taught graduate computing studies in the U.S. and abroad. He has held management, technical and advisory positions with GTE Data Services, American Software and Computer Services, Saudi Aramco, EC Cubed, the Technical Resource Connection division of Perot Systems and IBM Global Services. He developed technology transition plans for clients served by these companies, including GE, American Express, MasterCard and American Airlines-Sabre. In addition to numerous articles and professional papers, he is an author of ten landmark books. Peter has delivered keynote talks and papers to professional conferences in America, Austria, Australia, Canada, China, The Netherlands, South Africa, Japan, United Arab Emirates, Saudi Arabia, Egypt, Bahrain, Germany, Britain, Italy and France. www.peterfingar.com

Companion Book

www.mkpress.com/cloud

Watch for forthcoming titles.

Meghan-Kiffer Press
Tampa, Florida, USA
www.mkpress.com
Innovation at the Intersection of Business and Technology

Companion Book

The strategy guide for CEOs, CIOs—and the Rest of Us.
www.mkpress.com/ECC

Meghan-Kiffer Press
Tampa, Florida, USA
www.mkpress.com
Innovation at the Intersection of Business and Technology

www.mkpress.com